PHASE DIAGRAMS OF MULTICOMPONENT SYSTEMS

IZOBRAZHENIE KHIMICHESKIKH SISTEM S LYUBYM CHISLOM KOMPONENTOV

ИЗОБРАЖЕНИЕ ХИМИЧЕСКИХ СИСТЕМ С ЛЮБЫМ ЧИСЛОМ КОМПОНЕНТОВ

A Special Research Report

PHASE DIAGRAMS OF MULTICOMPONENT SYSTEMS
Geometric Methods

Fanya Moiseevna Perel'man
Institute of General and Inorganic Chemistry
of the Academy of Sciences of the USSR

Translated from Russian by
David A. Paterson
Paisley College of Technology
Paisley, Scotland

Springer Science+Business Media, LLC 1966

First Printing–April 1966
Second Printing–February 1967

ISBN 978-1-4899-4772-7   ISBN 978-1-4899-4770-3 (eBook)
DOI 10.1007/978-1-4899-4770-3

The original Russian text was published by Nauka Press for
the N. S. Kurnakov Institute of General and Inorganic Chemistry
in Moscow in 1965

Фаня Моисеевна Перельман
Изображение химических систем с любым числом компонентов

Library of Congress Catalog Card Number 66-12705

© 1966 Springer Science+Business Media New York
Originally published by Consultants Bureau Enterprises, Inc in 1966.
Softcover reprint of the hardcover 1st edition 1966

227 W. 17th St., New York, N. Y. 10011

PREFACE TO THE AMERICAN EDITION

The present monograph deals with the special branch of chemical statics which arose at the end of the nineteenth and beginning of the twentieth centuries, to meet an urgent need for a detailed study of many substances of practical importance, with variable composition, particularly, alloys, glass, fluxes, and slags. The properties of these substances vary over wide limits, depending on the component ratio and general conditions of formation. Thus the ordinary methods of preparative chemistry — synthesis and analysis — which had proved so fruitful when applied to definite chemical compounds of constant composition, were unsuitable in these cases. New methods and techniques — new theoretical premises — were required. This led to the development of a new field, involving essentially the study of the chemical transformations taking place as a result of the interaction between different substances, from the change in their relative concentration under the influence of temperature, pressure, and other external factors. The originators included van't Hoff, Roozeboom, Le Chatelier, Roberts-Austen, and many others. They relied on the classical work of Gibbs on thermodynamic equilibrium and on the concepts of chemical systems, components, and phases, which he introduced.

The development of this field in the Soviet Union is indissolubly bound with the name of the outstanding Russian scientist N. S. Kurnakov. As early as 1904 he constructed an extremely accurate instrument for measuring the heats of phase changes — a recording pyrometer — more efficient than the similar instruments used at that time in other countries. In the period 1906-1908, with S. F. Zhemchuzhnyi, he used this instrument for the first study of metallic systems: the nickel—copper and gold—copper systems. The first phase diagrams of the extremely important silicate system formed by the oxides of calcium, aluminum, and silicon were published in the USA in 1915 by Rankin and Wright.

At the present time, throughout the world, more than 3000 different systems — metallic, silicate, and salt — have been studied. In this vast amount of work, the contribution made by Soviet scientists has been considerable, particularly in the study of more complex systems containing three, four, and five components. Studies in this field (at N. S. Kurnakov's suggestion, often referred to as physicochemical analysis) have in turn required the development of special methods of representation, suitably reflecting the relationship between the properties and composition of phases in systems with a large number of components. The most rational methods have been found to be those based on the concepts and ideas of multidimensional geometry.

Multicomponent systems are widespread in nature and in industry. Examples include rocks, minerals, and ores, sea water, and salt-lake brines. Many technical alloys and semiconducting materials, heat-transfer agents and electrolytes, enamels and glasses, complex mineral fertilizers, and catalysts are multicomponent systems. The study of systems containing these components is greatly facilitated when geometric methods are used to summarize the experimental data. Moreover, graphs are generally concise and readily understood, and permit quantitative calculations, interpolation, and extrapolation. It thus becomes possible to predict the properties of multicomponent compositions on the basis of data on lower constituent systems from the same starting materials, and this is of considerable practical importance.

It is of course necessary to take account of the difficulty involved in understanding the abstract mathematical concepts of multidimensional geometric figures, corresponding to algebraic equations with many unknowns. I have therefore tried to make the resulting representations as clear as possible, and the actual construction of the phase diagrams has been carried out in simple, readily understood fashion. In this, however, the limit has not been reached, and further improvements are possible and desirable.

In dedicating this preface to the compatriots of J. W. Gibbs, I shall be happy if my book will in some measure facilitate the development of studies in the field of multicomponent systems of various types, which are acquiring ever increasing importance in modern technology.

April, 1966   F. M. Perel'man

BIOGRAPHICAL NOTE

Fanya Moiseevna Perel'man was born in Volozhin in the Belorussian SSR in 1896. Since 1934 she has worked at the Institute of General and Inorganic Chemistry of the Academy of Sciences of the USSR. She has published four monographs and more than 70 papers in Soviet periodicals; these chiefly concern work related to the study of various complex chemical systems.

PREFACE

Geometrical methods have long been used in various branches of chemistry. They are particularly valuable in the study of more complex processes taking place in the presence of a large number of components and under the influence of a variety of external factors. With increase in the total number of independent variables defining the state of a system, the need to use the concepts and ideas of multidimensional geometry becomes more pressing. Thus four-dimensional figures are used to represent the composition of five-component systems. The simplest of these figures, suitable for the construction of the diagrams for quinary systems of different types, together with methods for constructing their optimum projections on the plane of a diagram, are discussed in the author's earlier work "Methods of Representing Multicomponent Systems. Five-Component Systems" (Moscow, Izd. Akad. Nauk SSSR, 1959).

The present monograph deals with analogous problems relating to systems with any number of components, no matter how large, involving metals or salts, with or without double decomposition reactions. The monograph consists of two parts and a brief introduction, which gives a rational classification of multicomponent systems (proposed by V. P. Radishchev) and describes the few graphical methods at present used in the study of these systems.

The first part provides the theoretical basis for the method of optimum projections which I have developed. Its essential feature is that it uses, for the representation of multicomponent systems, those planar projections of the multidimensional figures which permit quantitative calculations.

After a description of the general regular features governing the formation of this type of "optimum projection," systems containing one, two, three, etc., different anions with any number of cations are considered in turn. Their structures, i.e., the number of one-, two-, three-, etc., component systems making up the overall systems, are described, and multidimensional geometric figures suitable for their representation are given, together with the optimum planar projections of these figures.

A description is also given of the optimum projections of some four-dimensional figures in three-dimensional space; this makes it possible to represent these systems by means of models.

The second part gives specific examples illustrating the practical utilization of the method of optimum projections for the representation of systems of different types. Here particular attention is paid to the prediction of the properties of multicomponent chemical compositions from extremely complex systems which have not been studied experimentally.

In particular a method is given for determining provisionally the melting points of some six-component alloys of nickel, iron, chromium, manganese, copper, and cobalt, and of nickel, titanium, chromium, molybdenum, tungsten, and niobium.

Of more general significance is the section which gives a complete provisional isotherm for solubility in the river system at 25°C. This makes it possible to determine which salts will crystallize first during the evaporation of a given ground water, whose salt composition is known, and this may provide an extremely important property for the classification of natural waters.

The text then deals with problems related to the construction of provisional partial solubility or fusion diagrams for reciprocal systems with more than six components, involving a large number of double decomposition reactions.

I wish to thank Doctor of Chemical Sciences Professor V. Ya. Anosov and Doctor of Chemical Sciences Professor I. I. Kornilov for reading the manuscript and for valuable advice.

F. M. Perel'man

PUBLISHER'S NOTE

The following Soviet journals cited in this book are available in cover-to-cover translation:

Russian title	English title	Publisher
Doklady Akademii Nauk SSSR	Doklady Physical Chemistry	Consultants Bureau
	Soviet Physics — Doklady	American Institute of Physics
Zhurnal Neorganicheskoi Khimii	Russian Journal of Inorganic Chemistry	The Chemical Society

CONTENTS

INTRODUCTION

Before the work of V. P. Radishchev there was no rigorous classification of chemical systems. Systems were characterized by the total number of components.

Thus all systems were divided into binary, ternary, quaternary, quinary, and multicomponent systems; at the same time a distinction was made between simple systems, in which no double decomposition or displacement takes place, and reciprocal systems, in which these reactions are observed.

This classification, however, was found to be clearly inadequate. A. G. Bergman was the first to point out that the term "reciprocal" is not definite, since reciprocal systems $C//A$ * with the same number of components may differ sharply in character [1, 2]. The simplest example is provided by five-component reciprocal systems, which may be of two fundamentally different types: 1) $4//2$ or $2//4$ — formed by four cations and two anions or from two cations and four anions; and 2) $3//3$ — formed by three cations and three anions.

With increase in the total number of components, the number of different possible types increases considerably and for nine-component reciprocal systems the following four types are possible: $8//2, 7//3, 6//4,$ and $5//5$.

The term "simple" likewise cannot be considered entirely appropriate, since in these systems (as in reciprocal systems) the components often react with one another to form a variety of more complex chemical compounds: crystal hydrates, double and triple salts, or complex compounds. V. P. Radishchev's idea was that chemical systems should be divided into various classes, depending on the number of ions of a given sign. In other words, all systems with the same number of anions and any number of cations (or with the same number of cations and any number of anions) should be included in the same class. Within each class the systems are further subdivided into types, depending on the total number of components [3-5].

When the systems of all possible classes from the first to, for example, the ninth are arranged in a rectangular table, it is possible to observe certain regular and common features in the properties of the systems of a given class (Table 1). The rectangle formed by the table is divided, by means of an imaginary diagonal

TABLE 1. Different Classes of Multicomponent Systems $(C//A)$

No. of anions A	Number of cations C								
	1	2	3	4	5	6	7	8	9
1	1//1	2//1	3//1	4//1	5//1	6//1	7//1	8//1	9//1
2	1//2	2//2	3//2	4//2	5//2	6//2	7//2	8//2	9//2
3	1//3	2//3	3//3	4//3	5//3	6//3	7//3	8//3	9//3
4	1//4	2//4	3//4	4//4	5//4	6//4	7//4	8//4	9//4
5	1//5	2//5	3//5	4//5	5//5	6//5	7//5	8//5	9//5
6	1//6	2//6	3//6	4//6	5//6	6//6	7//6	8//6	9//6
7	1//7	2//7	3//7	4//7	5//7	6//7	7//7	8//7	9//7
8	1//8	2//8	3//8	4//8	5//8	6//8	7//8	8//8	9//8
9	1//9	2//9	3//9	4//9	5//9	6//9	7//9	8//9	9//9

*Here and subsequently the symbol $C//A$ denotes a system formed by C cations and A anions.

TABLE 2. Degree of Interaction $B/N = R$ in Multicomponent Systems

No. of anions A	Number of cations C								
	1	2	3	4	5	6	7	8	9
1	$\frac{0}{1}=0$	$\frac{0}{2}=0$	$\frac{0}{3}=0$	$\frac{0}{4}=0$	$\frac{0}{5}=0$	$\frac{0}{6}=0$	$\frac{0}{7}=0$	$\frac{0}{8}=0$	$\frac{0}{9}=0$
2	$\frac{0}{2}=0$	$\frac{1}{4}=0.25$	$\frac{3}{6}=0.5$	$\frac{6}{8}=0.75$	$\frac{10}{10}=1.0$	$\frac{15}{12}=1.25$	$\frac{21}{14}=1.5$	$\frac{28}{16}=1.75$	$\frac{36}{18}=2.0$
3	$\frac{0}{3}=0$	$\frac{3}{6}=0.5$	$\frac{9}{9}=1.0$	$\frac{18}{12}=1.5$	$\frac{30}{15}=2.0$	$\frac{45}{18}=2.5$	$\frac{63}{21}=3.0$	$\frac{84}{24}=3.5$	$\frac{108}{27}=4.0$
4	$\frac{0}{4}=0$	$\frac{6}{8}=0.75$	$\frac{18}{12}=1.5$	$\frac{36}{16}=2.25$	$\frac{60}{20}=3.0$	$\frac{90}{24}=3.75$	$\frac{126}{28}=4.5$	$\frac{168}{32}=5.25$	$\frac{216}{36}=6.0$
5	$\frac{0}{5}=0$	$\frac{10}{10}=1.0$	$\frac{30}{15}=2.0$	$\frac{60}{20}=3.0$	$\frac{100}{25}=4.0$	$\frac{150}{30}=5.0$	$\frac{210}{35}=6.0$	$\frac{280}{40}=7.0$	$\frac{360}{45}=8.0$
6	$\frac{0}{6}=0$	$\frac{15}{12}=1.25$	$\frac{45}{18}=2.5$	$\frac{90}{24}=3.75$	$\frac{150}{30}=5.0$	$\frac{225}{36}=6.25$	$\frac{315}{42}=7.5$	$\frac{420}{48}=8.75$	$\frac{540}{54}=10.0$
7	$\frac{0}{7}=0$	$\frac{21}{14}=1.5$	$\frac{63}{21}=3.0$	$\frac{126}{28}=4.5$	$\frac{210}{35}=6.0$	$\frac{315}{42}=7.5$	$\frac{441}{49}=9.0$	$\frac{588}{56}=10.5$	$\frac{756}{63}=12.0$
8	$\frac{0}{8}=0$	$\frac{28}{16}=1.75$	$\frac{84}{24}=3.5$	$\frac{168}{32}=5.25$	$\frac{280}{40}=7.0$	$\frac{420}{48}=8.75$	$\frac{588}{56}=10.5$	$\frac{784}{64}=12.25$	$\frac{1008}{72}=14$
9	$\frac{0}{9}=0$	$\frac{36}{18}=2.0$	$\frac{108}{27}=4.0$	$\frac{216}{36}=6.0$	$\frac{360}{45}=8.0$	$\frac{540}{54}=10.0$	$\frac{756}{63}=12.0$	$\frac{1008}{72}=14$	$\frac{1296}{81}=16$

Note. The number of reciprocal salt pairs is calculated from the formula $B = C_C^2 \cdot C_A^2$, i.e., the product of the numbers of combinations from the total number of cations and from the total number of anions, taken in pairs. The number of simple salts is $N = CA$, i.e., it is equal to the product of the number of cations and the number of anions. $C_m^n = m!/[n!(m-n)!]$ where $n!$ represents factorial m, i.e., the product of the natural series of numbers from 1 to m, inclusive.

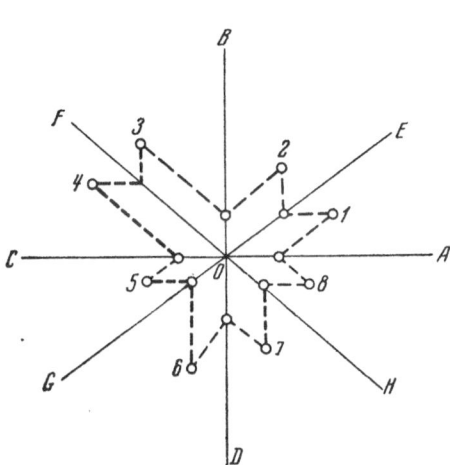

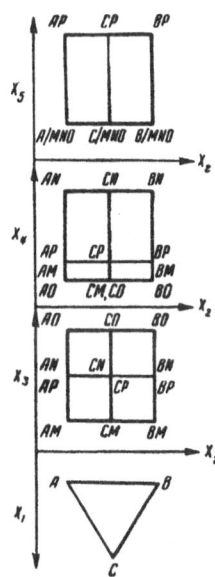

Fig. 1. Projections for the representation of the nine-component system of the first class ABCDEFGHO by Eitel's method. 1-8) Projections of the same point on the coordinate planes

Fig. 2. Projections for the representation of a six-component reciprocal system of the third class ABC//MNOP by Radishchev's method.

from the upper left-hand corner to the lower right-hand corner, into two sections, in each of which all types of system from the first class to the ninth, inclusive, are represented. Systems with the same number of anions are arranged in horizontal rows in the right-hand (upper) triangle, and systems with the same number of cations are arranged in vertical rows in the left-hand (lower) triangle. The diagonal itself passes through the systems, of all nine classes, in which the number of cations is equal to the number of anions. Since the two triangles are symmetrical, we can restrict ourselves to one only, for example, the upper right-hand triangle, and include with it the above imaginary diagonal.

Thus each row of the table gives systems of the first, second, third, etc., classes up to and including the ninth.

The systems usually described as simple obviously make up the first class. The remaining classes cover the reciprocal systems: the second class those with two anions, the third class those with three anions, the fourth class those with four anions, etc.

The diagonal intersecting the rectangle in Table 1 from the lower left-hand corner to the upper right-hand corner, and also all the lines parallel to this diagonal, cover types of systems in different classes but with the same number of components.

If in each square of Table 1 we indicate any property of the corresponding types of system, for example, their structure, which is characterized by the number of reciprocal salt pairs B and the total number of simple salts N, or by the ratio of these two quantities, which V. P. Radishchev called the degree of interaction R, the differences between the individual classes of system emerge particularly clearly. The values of $B:N=R$ are given in Table 2.

Comparison of Tables 1 and 2 leads to the following conclusions.

As the class of the systems becomes higher, the complexity of their structures increases rapidly.

Let us take for example systems of six components. In systems of the first class (6//1) there are six simple salts (components), and since there are no reciprocal salt pairs, the degree of interaction for these systems is equal to zero. Systems of the second class with the same total number of components (5//2) contain ten simple salts (here, as in all reciprocal systems, the total number of simple salts, or one-component systems, is greater than the number of components) and ten reciprocal salt pairs, so that the degree of interaction is equal to unity. Finally, in systems of the third class (4//3), there are 12 simple salts and 18 reciprocal pairs, so that R is equal to 1.5. Nine-component systems may belong to the first, second, third, fourth, and fifth classes. Here the number of simple salts increases from 9 to 25, the number of reciprocal salt pairs from 0 to 100, and the degree of interaction from 0 to 4.0, respectively. With further increase in the total number of components, the complexity of the structure of the systems increases even more rapidly. Thus, for example, in fifteen-component systems of the eighth class (8//8) there are 64 simple salts and 784 reciprocal salt pairs, and the degree of interaction reaches 12.25.

On the other hand, systems with different numbers of components but of the same class have similar structures. For example, with increase in the number of components by unity, the total number of simple salts increases by 2 in all systems of the second class, by 3 in systems of the third class, by 4 in systems of the fourth class, etc. The degree of interaction increases with each component in analogous fashion: by 0.25 for systems of the second class, by 0.5 for systems of the third class, by 0.75 for systems of the fourth class, by 1 for systems of the fifth class, by 1.25 for systems of the sixth class, etc.

Thanks to the similarity in structure, it is possible to represent all systems of the same class by geometrical figures with analogous structures. Increase in the number of components requires merely an increase in the dimensionality of the corresponding figure.

The problem of the representation of systems with a large number of components is not however, limited to the choice of the appropriate geometric figure. The methods of representing five-component systems were described in detail earlier [6]. A variety of methods also exists for the representation of systems with more than five components.

Eitel's method is a further development of the Boeke—Skowt method for quinary systems. The following are its essential features.

The sum of the concentrations of all n components of the system is taken as equal to a constant quantity, for example, 100%. This leaves n − 1 independent variables. The variables are plotted in pairs in different sectors around the origin of coordinates; there are (n − 1) coordinate axes, intersecting at a common point (the center). In Fig. 1 the nine-component system ABCDEFGHO is represented according to Eitel's method. The number of coordinate axes is equal to 8, and they lie on four straight lines intersecting at the origin of coordinates. The four quadrants of the Boeke—Skowt method are here replaced by eight sectors formed by each pair of coordinate axes, which in this case intersect at an angle of 45°.

A disadvantage of this method is that the system is represented in disconnected fashion. Moreover, in the case of reciprocal systems the choice of independent variables (components) is arbitrary, and a very large number of diagrams is required in order to represent all possible combinations. For the simple systems also, the number of diagrams, shown in Fig. 1, is almost a minimum, since for a completely clear representation of the interrelations in the system it is important to know the interrelations between each component and each of the others, and not merely the interrelations between arbitrarily selected pairs. The given case of the system ABCDEFGHO requires not 8 but 36 diagrams (i.e., the number of possible combinations from 9 components, taken in pairs). In spite of this, Eitel's method gives an adequate representation of the system and is suitable for quantitative calculations.

Radishchev's method was designed to give a clear representation of a system not in disconnected fashion but as a whole; V. P. Radishchev, unlike Boeke and Eitel, selected a particular disposition of the geometrical figure representing the compositions of the system, relative to the coordinate axes. Whereas in Eitel's method the coordinate axes are identified with the components of the system, in Radishchev's method there is a fundamental difference between them, and the system of coordinates exists independently. Radishchev's method, put forward for the representation of multicomponent systems, is a further development of his method for systems with five components. Thus in the appropriate multidimensional figure most suitable for the representation of the compositions of a given specific system, one of the triangular faces is superposed on one of the coordinate planes in such a way that the coordinates of the vertices of this face can conveniently be calculated from the diagram. From a knowledge of the geometric structure of the selected multidimensional figure, the coordinates of all its other vertices can then be readily calculated using distance formulas. The coordinates of the vertices, found in this way, are used to construct the projection of the figure on different coordinate planes.

V. P. Radishchev used this method to determine the projections on several coordinate planes for five- and six-dimensional figures representing six- and seven-component systems, respectively [8] (Fig. 2). He considers systems of different classes (first, second, third, and fourth), including not only anhydrous systems but also systems with one, two, or three solvents. V. P. Radishchev pointed out a method of representing systems with any number of components.

For practical purposes he considered it necessary to compare several projections of the figure on different coordinate planes. In the projection process, however, unequal contraction of the elements of the figure, with superposition of its parts, is unavoidable. Thus when an appropriate chemical system is represented by means of these projections, the concentrations of the components of the system in most cases are given on different scales, and this makes the utilization of such diagrams extremely difficult. Moreover, the ranges of crystallization of different phases of the system screen one another, making quantitative calculations impossible.

It was found that the projections of multidimensional figures on different coordinate places are not equivalent from the viewpoint of practical suitability for the construction of the phase diagrams for chemical systems. An analysis of the projections of different four-dimensional figures, obtained by Radishchev's method, showed that some of them have projections, on the coordinate planes, whose utilization does not require representation of the components on different scales, and in the projection process those parts of the figure which correspond to the ranges of crystallization of identical phases of the system coincide. These projections were called optimum projections [6]. It was established that the method of optimum projections is the most complete and, as will be described in detail later, permits not only the representation of systems, with any number of components, which have already been studied, but also the construction of provisional phase diagrams (or composition—property diagrams) for systems on the basis of data for their lower component systems. It should however be pointed out that other methods for the representation of multicomponent systems are possible in addition to those based on the use of multicomponent geometric figures and their projections.

In the study of simple and reciprocal quaternary systems, extensive use is made of Jänecke's method. The basis of this method is that the sum of the concentrations of any three components is taken as 100%, and the concentration of the fourth component is also expressed as a percentage of this sum [9, 10]. This method can in principle be extended to systems with any number of components [11].

The sum of the concentrations of any three components having been taken as 100%, the concentrations of each of the others are expressed as percentages of this sum. Thus the system is represented by means of a series of diagrams, each of which shows the selected three components (the sum of whose concentrations was taken as 100%) and a fourth component, whose percentage concentration relative to these three is given in the form of a separate projection or in the form of isoconcentrates, analogous to isotherms.

A second method involves the representation of a multicomponent system by means of sections of a figure representing the given system. The sections are drawn horizontally or vertically in such a way that the concentration of any one component or the ratio of the concentrations of any two components of the system remains constant.

The problem reduces to the representation of three independent variables, and this is readily done on the plane of a diagram [12, 13]. If the concentrations of any three components of the system are varied arbitrarily while the relative concentration of all the others is kept constant, the results of the study can be represented in the form of a function of four independent variables, by means of a model [14].

PART I

THE REPRESENTATION OF MULTICOMPONENT SYSTEMS
BY THE METHOD OF OPTIMUM PROJECTIONS

CHARACTERISTIC FEATURES OF THE OPTIMUM PROJECTIONS
OF FOUR-DIMENSIONAL FIGURES ON COORDINATE PLANES

Analysis of the planar projections of some four-dimensional figures whose formation was described in detail earlier [6] leads to the following conclusions.

Projections of the same geometric figure on different coordinate planes (for any chosen disposition of the figure relative to the coordinate axes) differ in their degree of clarity and practical suitability for the representation of the corresponding quinary systems. These differences are related to the direction of the projecting rays relative to the edges and faces of the figure. The following types of projection are possible.

First Type. Projection Rays Not Parallel to Any of the Edges of the Figure

As shown by B. N. Delone [15], in this case each vertex of the four-dimensional figure will be represented in three-dimensional space by a vertex, each edge by an edge, each face by a face, and each three-dimensional cell by a three-dimensional cell. Thus all the elements of the four-dimensional figure will be reflected in its projection. All its edges, however (with the exception of those perpendicular to the projection rays), will undergo contraction, to different extents. Thus all the faces and cells including those contracted edges will also undergo contraction. The dimensionality of the original figure will obviously be reduced by one, and the projection of the four-dimensional figure will be a three-dimensional model. The latter can be projected on to the plane of a diagram in such a way that again the projection rays are not parallel to any of the edges. We then obtain a planar representation which retains completely all the vertices, edges, and faces of the model, although these will have undergone further contraction. The three-dimensional cells and the model as a whole will be represented by corresponding planar figures (Figs. 3a, b, and c).

Since the number of rays not parallel to a given direction may be infinitely large, the number of such models and their planar representations for each four-dimensional figure will also be infinitely large. All such projections, however, will belong to the same type, with the following characteristic features.

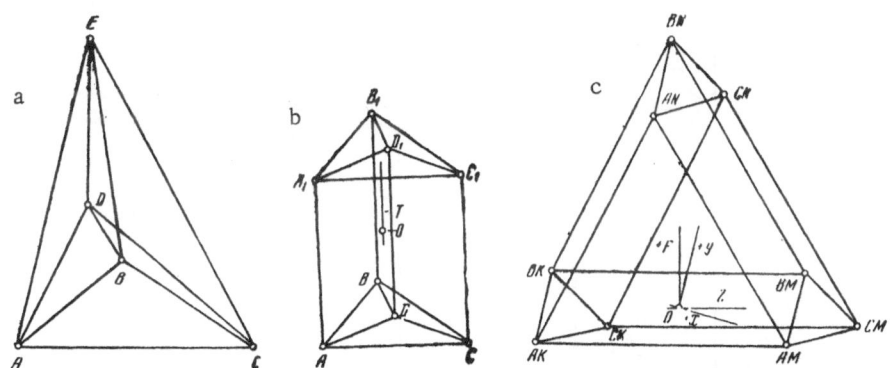

Fig. 3. Projections of the first type on the plane of a diagram. a) Pentatope; b) tetrahedral hexahedroid; c) prismatic hexahedroid.

9

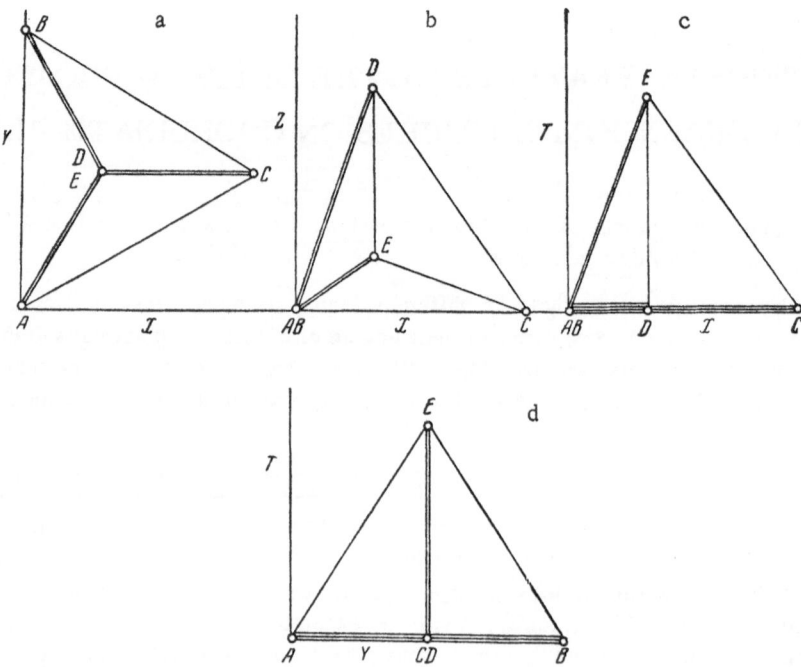

Fig. 4. Projections of a pentatope, of the second type (a–d), on different coordinate planes.

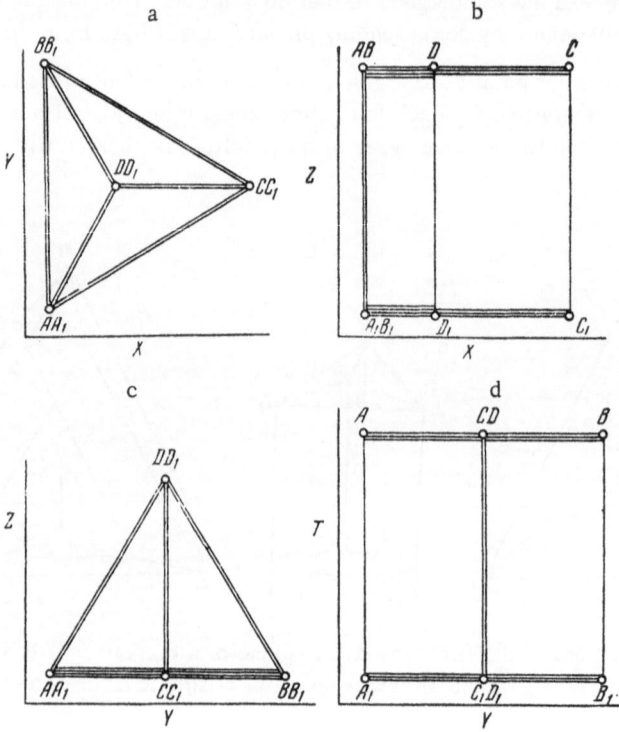

Fig. 5. Projections of a tetrahedral hexahedroid, of the second type (a–d), on different coordinate planes.

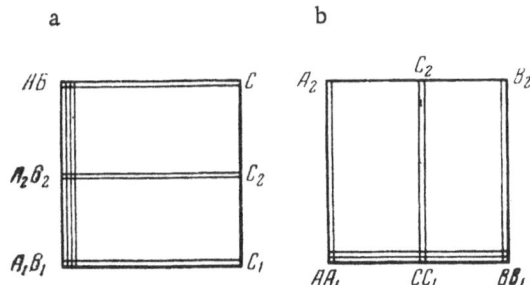

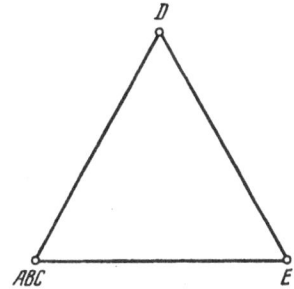

Fig. 6. Projections of a prismatic hexahedroid, of the second type (a, b), on different coordinate planes.

Fig. 7. Planar projection of a pentatope, of the third type.

In the first place, projections of the first type make it possible to obtain information on the structure of the system which is represented by the corresponding figure, i.e., on the number and nature of the one-, two-, three-, etc., component systems making up the system; in the second place, these projections make possible a qualitative estimation of the structure of the system, i.e., the presence in the system of various phases formed at different component ratios. Projections of the first type are quite unsuitable, however, for quantitative calculations.

Second Type. Projection Rays Parallel to One of the Edges of the Four-Dimensional Figure (or to Several Edges Parallel to One Another)

It is obvious that the edge (or edges) to which the projection rays are parallel will undergo contraction to zero in the projection, and its two vertices will merge into one.

All the faces including this "parallel" edge are converted into straight lines, and all three-dimensional cells containing them are converted into planar faces.

The original four-dimensional figure, however, contains other edges, which are not parallel to the projection rays. All of these, and also the faces and cells formed only by these "nonparallel" edges, are reflected on the diagram in exactly the same way as in the case of the projections of the first type already described. Thus the projection of the four-dimensional figure in three-dimensional space, obtained under these conditions, will again be a three-dimensional model, but, unlike the case of the first model, individual elements of the original figure will be degenerate in the model. This model can again be represented on a plane; in this second projection the projection rays should not be parallel to any of the edges of the model. Figures 4, 5, and 6 show some projections of the second type.

If we transfer our attention from the four-dimensional figures to the quinary systems which are represented by means of these projections, the following feature is observed. For systems of the first class consisting of five components, projection 4b is slightly more convenient, although in it only three of the components of the system are represented; in this case it is possible to make a more correct estimate of the approximate boundaries of the ranges of crystallization of most of the phases of the system.

For five-component systems of the second class, suitable projections of the second type are projections 5b and 5d. In these cases, not all of the eight simple salts of the original system are degenerate and represented in pairs, but only four of them. It is therefore possible, if only qualitatively, to represent the other four salts and also the ranges of crystallization of phases rich in these salts.

Finally, for quinary systems of the third class, both of the available (essentially identical) projections of the second type characterize the approximate boundaries of the ranges of crystallization of the phases rich in three simple salts of the system (Figs. 6a and b).

All projections of the second type are unsuitable for quantitative calculations, since the contraction is different for different elements of the figure, and also, more particularly, since its separate parts reflecting the various components and parts of the system are superimposed.

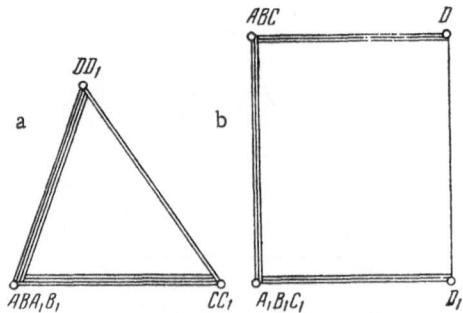

Fig 8. Planar projections of a tetrahedral hexahedroid, of the third type.

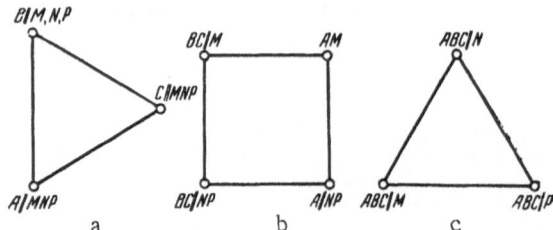

Fig. 9. Planar projections of a prismatic hexahedroid, of the third type.

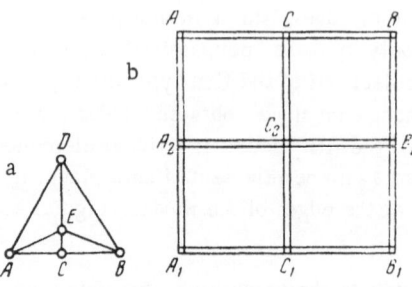

Fig. 10. Planar projections of the first type, with degenerate faces: a) pentatope; b) prismatic hexahedroid.

Third Type. Projection Rays Parallel to One of the Faces of the Figure or to Several Faces Parallel to One Another

Since the face as a whole, and hence each of the edges forming it, are parallel to the projection rays, all the vertices of these edges merge on the projection into one vertex, and all the edges and the face itself degenerate into a single point. All the faces adjoining this "parallel" face are converted into straight lines, since one of their sides (in the case of triangular faces) or two parallel sides (in the case of square faces) are converted into points. All the cells which contained this face which degenerated into a point are converted into planar faces.

In the general case, however, the original four-dimensional figures may contain other faces and other cells, which do not share edges with the "parallel" face. These therefore do not contain edges parallel to the projection rays, and are represented in the projection in all their elements, merely undergoing a certain contraction.

As a result, the projections of four-dimensional figures of the third type are also three-dimensional models, although the degeneracy of the individual elements is much greater than in the previous case. Further projection of this model by means of rays which are not parallel to any of its edges gives a planar projection which reflects completely all the geometric elements of this model (Figs. 7, 8, and 9).

In order to estimate the properties of projections of the third type, let us examine the application of these figures to the problem of representing corresponding quinary systems.

Figure 7 shows the projection of a pentatope on one of the coordinate planes; this projection is applicable for the representation of systems of the first class ABCDE. Three (ABC) of the five components of the system are represented together on the projection; the other two (D, E) are each represented separately. In addition, the superimposed edges of the original figure — AE, BE, and CE on the one hand and AD, BD, and CD on the other — undergo contraction to equal extents. As a result, we have on this projection superposition and complete coincidence of three adjacent faces of the pentatope, corresponding to the ternary systems AED, BED, and CED, with a common binary system. It is obvious that the superposition and coincidence of individual parts of the figure as a result of projection does not in this case lead to superposition and coincidence of the ranges of crystallization of phases of different kinds in the system close to the vertices E and D, since these parts of the diagram represent the ranges of crystallization of phases formed by the components E and D in all lower systems containing them (i.e., ternary and quaternary systems) and in the quinary system itself. Thus this projection can be used to construct phase diagrams or com-

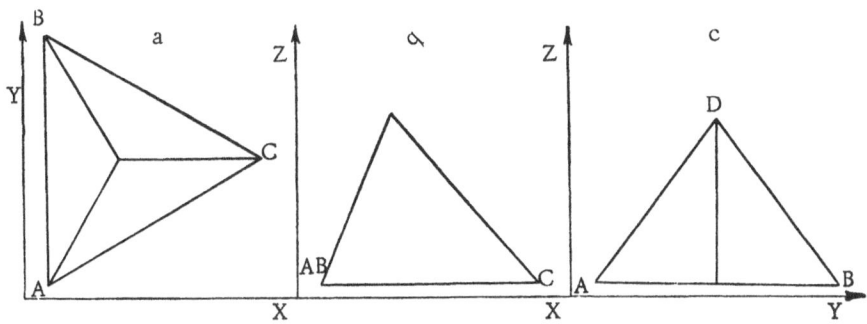

Fig. 11. Projections of a tetrahedron on the coordinate planes: a) XY; b) XZ; c) YZ.

position – property diagrams which make it possible to determine the boundaries of the ranges of crystallization of the separate phases of the system ABCDE and to carry out other quantitative calculations. Such projections are called optimum projections. Thus for the pentatope the optimum projection coincides with the
only projection of the third type for this figure. The situation is slightly more complicated for the case of
four-dimensional figures representing systems of other classes.

The tetrahedral hexahedroid, which is used to represent systems of the 4//2 type, has two projections of
the third type. One of these (Fig. 8a) is obtained by projection by means of rays parallel to one of the square
faces of the figure. Since two edges of this square face are parallel to two other edges of the hexahedroid, these
edges are also parallel to the projection rays and hence degenerate into points on the projection. In other words,
the projection obtained shows not only degeneracy of one of the faces into a point but also degeneracy of two
other edges of the four-dimensional figure, which are not part of the "parallel" face.

Since each vertex of the four-dimensional figure corresponds to one of the simple salts of the system, we
find that of the eight simple salts in the system, formed by combining each cation with each anion, four salts,
for example, A, B, A_1, and B_1, are represented jointly, since they correspond to the vertices of the "parallel"
face, whereas the four remaining salts are represented together in pairs, since D and D_1, and also C and C_1,
correspond to the vertices of two "parallel" edges. Thus none of the salts of the system is represented separately, all being represented together with some other salt.

The second projection of the third type (Fig. 8b) is obtained by projecting the tetrahedral hexahedroid
by means of rays which are parallel to two mutually-parallel triangular faces of the original figure. Thus six
of the simple salts of the system are represented jointly in threes at corresponding vertices of the projection.
The other two salts of the original system (D and D_1), however, are represented in Fig. 8b as individual vertices.
Since the coincident edges undergo contraction to the same extent and since in this case three adjacent faces
coincide, we have all the conditions necessary for the formation of an optimum projection.

Thus in the case of the tetrahedral hexahedroid, in order to obtain the optimum projection on a coordinate plane, it is necessary to carry out the projection by means of rays which must be parallel not merely to
any of its faces but specifically to one of its triangular faces. If this is not the case, the projection of the third
type is not the optimum projection.

The prismatic hexahedroid, used to represent quinary systems of the third class, has three planar projections of the third type. Two of these (Figs. 9a and c) are identical. These projections were obtained by
projecting the original figure by means of rays parallel to one of the triangular faces of the original figure.
Each of the six triangular faces of the prismatic hexahedroid is parallel to two others.

If the quinary system represented by the given figure is denoted by the symbol ABC//MNP, the triangular
faces correspond to six simple ternary systems from the same nine salts: three systems formed by salts with a
common anion: 1) AM–BM–CM; 2) AN–BN–CN; and 3) AP–BP–CP; and three systems formed by salts with
a common cation: 1) AM–AN–AP; 2) BM–BN–BP; and 3) CM–CN–CP. On the prismatic hexahedroid these
systems are represented in such a way that the corresponding three first and three second triangular faces are
mutually parallel.

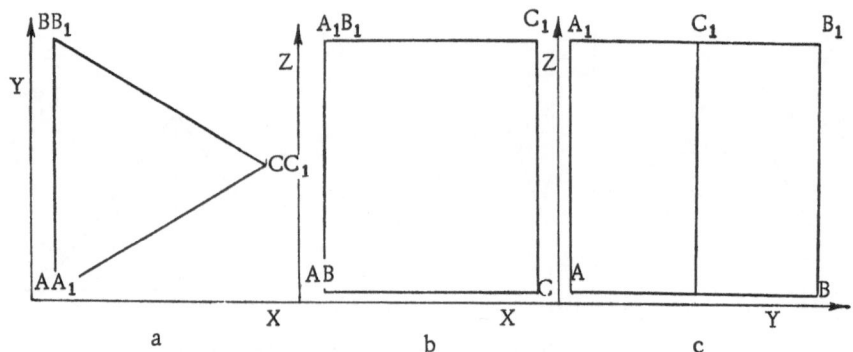

Fig. 12. Projections of a triangular prism on the coordinate planes: a) XY; b) XZ; c) YZ.

As a result, on these projections, three of the triangular faces of the original figure degenerate into a point; this makes it necessary, when these projections are used to construct the phase diagrams of the corresponding systems, to represent together salts with the same cation or with the same anion.

The projections in Figs. 9a and c are not optimum projections, since they do not reflect the ranges of crystallization of individual phases of the system. These projections, however, can be used for quantitative estimations in those cases where the study of the system involves the investigation of problems relating only to the anions or only to the cations.

Figure 9b, obtained by projecting the prismatic hexahedroid by means of rays parallel to one of the square faces of this figure, is distinguished by different properties, since none of the nine square faces of the figure is parallel to two others. Here, in addition to the degeneracy of the "parallel" face (BC//NP) into a point, we also have degeneracy into a point for two edges (A//NP and BC//M). One of the vertices of the hexahedroid (AM) is here reflected directly and does not merge with any other vertex of the original figure. Since each of the nine vertices of the hexahedroid corresponds to one of the nine simple salts of the system ABC//MNP, Fig. 9b makes it possible to represent quantitatively the boundaries of the ranges of crystallization of the phases formed by one of these salts and hence is an optimum projection.

Thus in order to obtain the optimum projection of a prismatic hexahedroid on a coordinate plane it is necessary to carry out the projection by means of rays parallel to one of its square faces. If the projection rays are parallel to one of the triangular faces, the projection obtained is suitable for the quantitative representation of the corresponding systems only in individual particular cases.

In the examination oi the planar projections of four-dimensional figures, we started from the assumption that these figures are first projected into three-dimensional space, after which the models obtained are projected again on to a plane. The projection on to a plane is always carried out by means of rays which are not parallel to any of the edges of the model obtained as a result of the first projection.

Another possibility is that the second projection be carried out by means of rays parallel to the height of the model; in a number of cases this is not parallel to a corresponding edge. The projection of the original four-dimensional figure may then lead to other types of projection (Figs. 10a and b).

It should also be noted that the nature of the projections on to different coordinate planes is determined exclusively by the adopted disposition of the original geometric figure relative to the system of coordinates. We have indicated only those projections which can be obtained with the most convenient orientation of the figure relative to the coordinate axes.

It is apparent that not only four-dimensional but also three-dimensional figures may have projections of different types, depending on the method of projection.

It is particularly important to know the optimum projections of two such figures — the tetrahedron and triangular prism — most frequently used in physicochemical analysis.

For each three-dimensional figure, three projections on coordinate planes are possible.* Of course, the optimum projection should in this case be a projection of the second type, i.e., one obtained by projection by means of rays parallel to one of the edges of the original figure. For the tetrahedron, representing quaternary systems of the first class, there is only one projection of the second type, and this is also an optimum projection (Fig. 11b). For the triangular prism, representing quaternary systems of the second class, two projections of the second type are possible. One of these is formed by projection by means of rays parallel to two edges which are part of triangular faces of the prism (Fig. 12b) and the other is formed by projection by means of rays parallel to edges forming part of all its square faces (Fig. 12a).

It can readily be seen that only the first projection is an optimum projection, the second being suitable only in individual particular cases.

---

*The number of coordinate planes is equal to the number of combinations taken in pairs from the total number of coordinate axes, which is equal to the dimensionality of the figure. For a three-dimensional figure we have $C_3^2 = 3$.

OPTIMUM PLANAR PROJECTIONS OF MULTIDIMENSIONAL FIGURES REPRESENTING SYSTEMS OF THE FIRST AND SECOND CLASSES

The Structure of Multicomponent Systems of the First Class

Systems of the first class C//1 have a comparatively simple structure. There is no double decomposition in these systems and the total number of one-component systems (or simple salts) is equal to the total number of components. All other lower constituent systems—binary, ternary, quaternary, etc. — also belong to the first class, and their number is determined as the number of combinations from the total number of components in pairs, threes, fours, etc. In general, in an n-component system, the number of $(n - k)$-component constituent systems is equal to C_n^{n-k}. As shown by N. S. Kurnakov [16], the structure of multicomponent systems of the first class can be determined by means of a slightly modified version of Pascal's arithmetic triangle (Table 3).

In Table 3, each row of numbers can be obtained from the previous one. For this purpose it is necessary merely to recognize that the figures in the second column form the natural series of numbers, while all the others are given by the sum of two terms which are numbers of the preceding horizontal row — one directly above the number being determined and the other in the preceding column.

The structure of multicomponent systems of the first class corresponds to the structure of the simplest multidimensional figures — regular simplexes.

The term "regular n-dimensional simplex" is used to describe the simplest closed convex n-dimensional figure defined by $(n+1)$ points with independent positions, i.e., not situated in the same $(n-1)$-dimensional space. One-component systems correspond to vertices, binary systems to edges, ternary systems to triangular faces, quaternary systems to tetrahedra, quinary systems to pentatopes, etc. Simultaneously the structure of the simplexes makes it possible to reflect the interrelationship between the components, for example the participa-

TABLE 3. Number of Lower Constituent Systems in Systems of the First Class

Symbol for system 1,2,3...,n components	Nature of system						
	One-component	Binary	Ternary	Quaternary	Quinary	Senary	Septenary
S_1	1	—	—	—	—	—	—
S_2	2	1	—	—	—	—	—
S_3	3	3	1	—	—	—	—
S_4	4	6	4	1	—	—	—
S_5	5	10	10	5	1	—	—
S_6	6	15	20	15	6	1	—
S_7	7	21	35	35	21	7	1
S_n^*	C_n^1	C_n^2	C_n^3	C_n^4	C_n^5	C_n^6	C_n^7

* The system S_n is bounded in the limit by S_{n-1}-component systems, the number of which is equal to n.

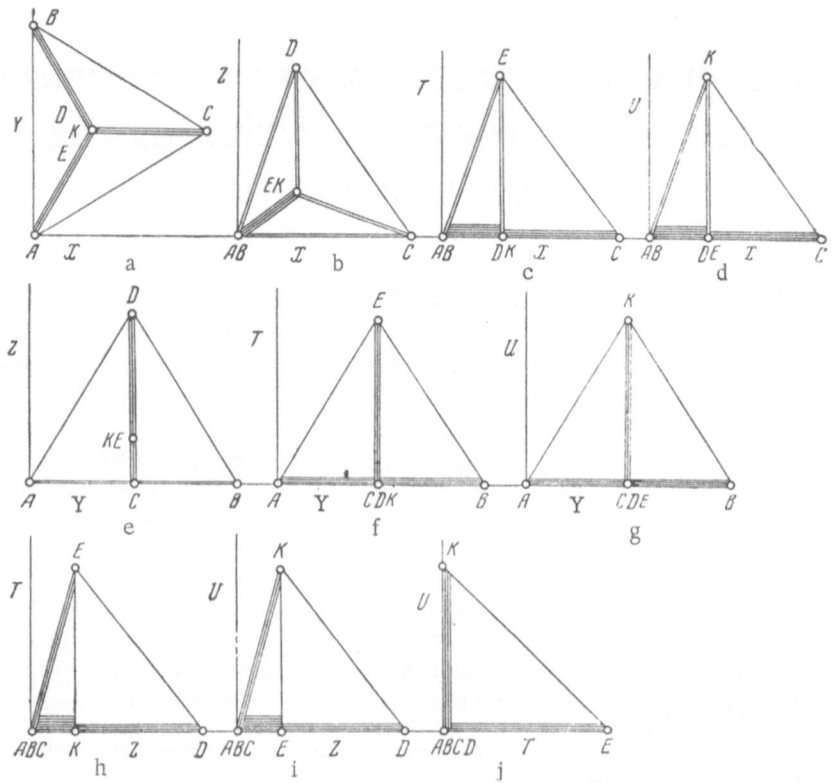

Fig. 13. Projections of a hexatope on ten coordinate planes: a) XY; b) XZ; c) XT;
d) XU; e) YZ; f) YT; g) YU; h) ZT; i) ZU; j) TU.

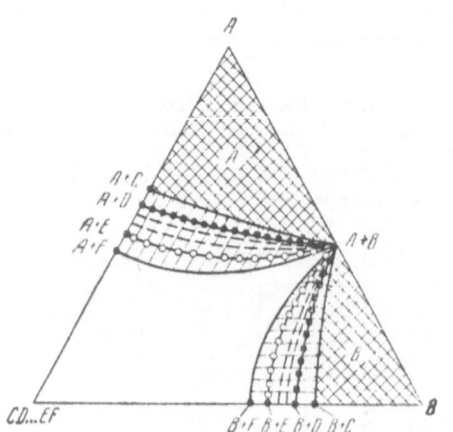

Fig. 14. Optimum projection of an n-dimensional regular simplex on the plane of a diagram.

tion of the same binary systems in the formation of three
different ternary systems forming part of a quinary system,
or of four different ternary systems forming part of a
senary system.

Optimum Projections of Simplexes

After the choice of the multidimensional figure
most suitable for the representation of the composition of
a multicomponent system has been made, there arises the
problem of projecting this figure on coordinate planes in
order to obtain graphs which make possible clear representations and quantitative calculations for the processes
taking place in the system, depending on its component
ratio and other equilibrium factors.

We can use the approach adopted for five-component systems: having positioned one of the faces of the
n-dimensional simplex in a particular way relative to the
system of coordinates, we calculate the coordinates of the vertices on the basis of geometric relationships or
from distance formulas. The values of the coordinates of the vertices can be used to construct projections on
all coordinate planes and to select the optimum projection from the planar projections obtained. The projections of a hexatope on ten coordinate planes, obtained in this way, are shown in Fig. 13.

The same result is achieved much more simply and rapidly if we use the general features governing the
formation of optimum projections for different classes of multidimensional figure.

TABLE 4. Structure of Systems of the Type 5//2, 6//2, 7//2, C//2, of the Second Class

Lower constituent systems	Formulas for calculating the total number of systems			
	5//2	6//2	7//2	C//2
One-component	$C_5^1 \cdot C_2^1$	$C_6^1 \cdot C_2^1$	$C_7^1 \cdot C_2^1$	$C_C^1 \cdot C_2^1$
Binary	$C_5^2 \cdot C_2^1 + C_5^1 \cdot C_2^2$	$C_6^2 \cdot C_2^1 + C_6^1 \cdot C_2^2$	$C_7^2 \cdot C_2^1 + C_7^1 \cdot C_2^2$	$C_C^2 \cdot C_2^1 + C_C^1 \cdot C_2^2$
Ternary	$C_5^3 \cdot C_2^1 + C_5^2 \cdot C_2^2$	$C_6^3 \cdot C_2^1 + C_6^2 \cdot C_2^2$	$C_7^3 \cdot C_2^1 + C_7^2 \cdot C_2^2$	$C_C^3 \cdot C_2^1 + C_C^2 \cdot C_2^2$
Quaternary	$C_5^4 \cdot C_2^1 + C_5^3 \cdot C_2^2$	$C_6^4 \cdot C_2^1 + C_6^3 \cdot C_2^2$	$C_7^4 \cdot C_2^1 + C_7^3 \cdot C_2^2$	$C_C^4 \cdot C_2^1 + C_C^3 \cdot C_2^2$
Quinary	$C_5^5 \cdot C_2^1 + C_5^4 \cdot C_2^2$	$C_6^5 \cdot C_2^1 + C_6^4 \cdot C_2^2$	$C_7^5 \cdot C_2^1 + C_7^4 \cdot C_2^2$	$C_C^5 \cdot C_2^1 + C_C^4 \cdot C_2^2$
Senary	$C_5^5 \cdot C_2^2$	$C_6^6 \cdot C_2^1 + C_6^5 \cdot C_2^2$	$C_7^6 \cdot C_2^1 + C_7^5 \cdot C_2^2$	$C_C^6 \cdot C_2^1 + C_C^5 \cdot C_2^2$
Septenary		$C_6^6 \cdot C_2^2$	$C_7^7 \cdot C_2^1 + C_7^6 \cdot C_2^2$	$C_C^7 \cdot C_2^1 + C_C^6 \cdot C_2^2$
C-component				$C_C^C \cdot C_2^1 + C_C^{C-1} \cdot C_2^2$

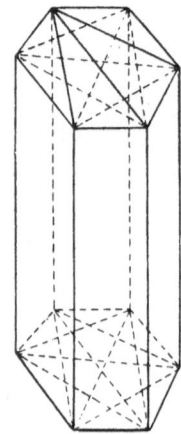

Fig. 15. Six-dimensional geometric figure for the representation of seven-component systems of the second class 6//2.

It was shown that the optimum projection of a pentatope on coordinate planes is obtained as a projection of the third type, i.e., by projection by means of rays parallel to one of its faces. It can readily be seen from Fig. 13 that the optimum planar projection of a hexatope is formed as a projection of the fourth type, i.e., when the projection is carried out by means of rays parallel to one of its tetrahedra (in Fig. 13j, the tetrahedron ABCD is the "parallel" one).

We thus arrive at the general conclusion: in order to obtain the optimum planar projection of a $(C-1)$-dimensional simplex representing a C-component system of the first class $C//1$, it is necessary to project the figure by means of rays parallel to one of its $(C-3)$-dimensional faces. From the above figures it also follows that for each simplex, only one of the projections on the coordinate planes is an optimum projection. The total number of graphs (constructed as the optimum-projection type) necessary for the complete representation of multicomponent systems of the first class is not less than $C//2$, where C is the number of components (when C is odd, the fraction is equated to unity).

The general form of an optimum projection of this type on the plane of a diagram is shown in Fig. 14.

If the simplex represents the C-component system ABCD...EF, Fig. 14 shows the minimum and maximum boundaries of the ranges of crystallization of phases containing components A and B, separately or together.

The Structure of Multicomponent Systems of the Second Class [17, 18]

The second class includes the reciprocal systems $C//2$, i.e., formed by two anions with any number of cations (or $2//A$, formed by two cations with any number of anions). Since each cation can combine with each of the two anions to form a simple salt, the total number of original simple salts of the system is greater than the total number of its components. If we take the ions as variables and make use of the fact that the total molecular concentrations of all the cations are equal to the corresponding total concentrations of all the anions, the total number of components is here equal to $C+2-1 = C+1$; at the same time the total number of original simple salts is $C_C^1 \cdot C_2^1 = C \cdot 2 = 2C$.

In order to determine the structure of the system $C//2$, i.e., to determine the nature and number of binary, ternary, quaternary, etc., systems present in it, let us examine in more detail the six-, seven-, and eight-component systems of this class, i.e., systems of the type $5//2$, $6//2$, and $7//2$.

The numbers of simple salts or one-component systems for the above systems are 10, 12, and 14, respectively. The formation of binary systems requires the combination of any two cations with each of the anions or of both anions with each of the cation.

Ternary systems may be either simple (i.e., of the first class) or reciprocal (i.e., of the second class).

Simple ternary systems are obtained by the combination of each set of three cations with one of the anions, and reciprocal ternary systems are obtained by the combination of each pair of cations with both anions.

Systems of four, five, and more components may also belong to these two classes (first and second), since it is evident that reciprocal lower systems cannot belong to other classes (apart from the second), as the original system contains only two anions. Here systems of the first class are formed by the combination of four, five, or more cations, respectively, with each of the anions, and reciprocal systems are formed by the combination of each set of three, four, or more cations, respectively, with both anions. On the basis of these general principles we can compare the data on the structures of these types of system of the second class, as shown in Table 4.

The table shows clearly the common character of the structures of all systems of the second class and the regular features of the change in structure with increase in the total number of components by unity. Systems of the second class have a much more complex structure than simple systems.

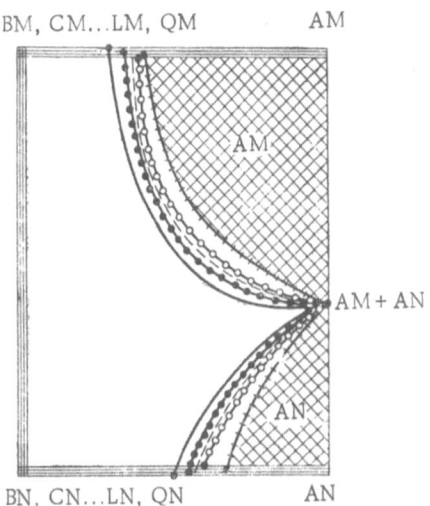

Fig. 16. Optimum projection of n-dimensional figures representing systems of the second class, on the plane of a diagram.

In systems C//2 the total numbers of lower constituent systems are expressed (apart from the case of one-component systems) by means of pairs of terms, the second members of which form a Pascal's triangle, while the first members form a doubled Pascal's triangle, since $C_2^2 = 1$ and $C_2^1 = 2$.

From this it follows that multidimensional simplexes cannot be used to represent systems of the second class. For unique correspondence between the structure of a system and the structure of the geometric figure it is obviously necessary that the number of one-component systems formed be equal to the number of vertices of this figure.

It is known that quaternary systems of the second class (of the type 3//2) are best represented by a triangular prism, and quinary systems (of the type 4//2) by a tetrahedral hexahedroid. In both cases the two anions of the system correspond to a straight line (the height of the prism or hexahedroid), the bases being triangles or tetrahedra, respectively. Thus the bases are simplexes for which the number of vertices corresponds to the number of cations in the system. From this it follows that with increase in the number of components it is necessary merely to increase the dimensionality of the simplexes forming the base of the figure, while retaining its general prismatic character. In other words, for the representation of systems of the second class it is necessary to use geometric figures of the required dimensionality, analogous to a tetrahedral hexahedroid (Fig. 15).

These figures are sections of simplexes with higher dimensionality and can in turn be sectioned into simplexes. Jänecke showed that a three-dimensional triangular prism is sectioned by means of two triangular sections to give three tetrahedra. Radishchev established that a four-dimensional tetrahedral hexahedroid is sectioned by three tetrahedra to give four pentatopes [19]. From this it follows that the analogous five-dimensional figure can be broken down by means of four pentatopes into five hexatopes, and that in general an n-dimensional figure of this type can be broken down by means of (n − 1) regular (n − 1)-dimensional simplexes to give n n-dimensional simplexes.* It should be noted that if we wished to represent a system of the second class C//2 by means of simplexes, each of these simplexes would correspond to a fraction equal to approximately 1/C of the system. Inconvenient features in this case include not only the large number of figures but also the fact that in the phase diagrams it would be necessary to take account of the possibility that the ranges of crystallization of the phase formed by (C − 1) components might extend beyond the limits of the corresponding simplex. In view of the reversibility of double decomposition reactions, this extremely probable situation would create additional difficulties.

Thus multidimensional figures analogous to a tetrahedral hexahedroid are most suitable for the representation of systems of the second class. When they are used, the sum of the concentrations of both anions is taken as 100% and the sum of the concentrations of all the cations is also taken as 100%.

Figure 15 shows a model of a six-dimensional figure,† suitable for the representation of seven-component systems of the type 6//2. It consists of two five-dimensional hexatopes, whose corresponding vertices are joined by straight lines in the sixth dimension. The vertices and edges of the figure correspond to one-component and two-component systems, and the faces and three-dimensional, four-dimensional, and five-dimensional cells correspond, respectively, to the ternary, quaternary, quinary, and senary systems, simple and reciprocal, present in the original seven-component system. To verify the accuracy of the formulas given in

*The simplexes obtained here, however, are irregular.

†In this case and below, the n-dimensional geometric figures are represented in the form of projections of the first type.

Table 4, let us use them to calculate the number of lower constituent systems. Since the corresponding multi-dimensional figures have exactly the same number of geometric elements — vertices, edges, faces, etc. — the data obtained should satisfy the Euler—Poincaré formula:

$$P_0 - P_1 + P_2 - P_3 + P_4 - \ldots \pm P_{n-1} = 1 - (-1)^n,$$

where P with subscript is the number of geometric elements of figures with zero, one, two, etc., dimensions and n is the dimensionality of the original figure.

We obtain the following results:

for systems $5//2$

$$10 - 25 + 30 - 20 + 7 = 1 - (-1)^5, \text{ i.e., } 2 = 2;$$

for systems $6//2$

$$12 - 36 + 55 - 50 + 27 - 8 = 1 - (-1)^6, \text{ i.e., } 0 = 0;$$

for systems $7//2$

$$14 - 49 + 91 - 105 + 77 - 35 + 9 = 1 - (-1)^7, \text{ i.e., } 2 = 2.$$

Agreement is obtained in all cases. Thus the data given in Table 4 for multicomponent systems of the second class are correct.

Optimum Projections of Figures Analogous to a Tetrahedral Hexahedroid

In order to determine the optimum projections of five-, six-, seven-, etc., dimensional figures — analogs of the tetrahedral hexahedroid — it is possible to adopt the approach used in the determination of the optimum projection of the four-dimensional figure itself. Having positioned the figure in suitable fashion relative to the system of coordinates, we next determine the values of the coordinates of the vertices, use these to construct projections of the figure on all the coordinate planes, and from the projections obtained select the optimum projection. We shall however adopt a simpler approach, based on the regular features of the formation of the optimum planar projection of the tetrahedral hexahedroid. It was shown above that for this four-dimensional figure the optimum projection on the coordinate planes is a projection of the third type, i.e., obtained by projection by means of rays parallel to one of the faces of the figure. The rays must be parallel to one of its triangular (not square) faces. Since the multidimensional analog-figures exhibit similar properties, we find that for geometric figures, with any dimensionality, representing systems of the second class, the optimum projection on the coordinate planes should be obtained with the same projection conditions, with allowance, of course, for the dimensionality of the figure. In other words, the projection rays should be parallel to simplexes including only triangular (and not square) faces of the original figure. It is evident that the dimensionality of this "parallel" figure should be smaller by 2 units than that of the original figure. Thus if the original C-dimensional figure represents a system $C//2$ formed by $(C + 1)$ components, each of the "parallel" figures is $(C - 2)$-dimensional, since they represent systems $(C - 1)//1$ formed by $(C - 1)$ components.

This planar projection has all the characteristics of an optimum projection (Fig. 16). Two of its vertices are represented separately, and these vertices correspond to two of the original simple salts of the system. Moreover, the coincident elements of the figure undergo contraction to equal extents. This makes it possible to represent all the components of the system on the same scale. Finally, the coincident parts of the figure correspond to the regions of crystallization of identical phases formed in the system, and this makes it possible to use this diagram for quantitative calculations.

Since one projection gives a clear representation of the regions of crystallization of the phases formed by only two of the simple salts, of which the total number in the system is 2C, the minimum number of planar projections of the optimum type, necessary for the representation of the system as a whole, is equal to C.

OPTIMUM PLANAR PROJECTIONS OF MULTIDIMENSIONAL
FIGURES REPRESENTING SYSTEMS OF THE THIRD CLASS

The Structure of Systems of the Third Class [20]

Systems of the third class $C//3$ contain three anions with any number of cations (or three cations with any nymber of anions $3//A$).

The simplest representatives of this class — five-component systems — are well known in the literature. They include, for example, the so-called river system Na^+, Ca^{2+}, $Mg^{2+}//Cl^-$, SO_4^{2-}, $HCO_3^- + H_2O$, if we restrict ourselves to its salt composition.

Let us examine the structure of six-, seven-, and eight-component systems of the types $4//3$, $5//3$, and $6//3$. The general principles of the calculation of the number of lower constituent systems is here based on the following considerations.

The number of one-component systems (simple salts) in all cases is equal to the product of the number of cations and the number of anions.

Binary systems are produced by the combination of each pair of cations with one of the anions and of each pair of anions with one of the cations: ternary simple systems are produced by the combination of each set of three cations with each anion and of all three anions with each cation, and ternary reciprocal systems by the combination of each pair of cations with each pair of anions.

In the same way, quaternary, quinary, etc., systems may also be either simple or reciprocal.

Simple quaternary systems, for example, are formed by the combination of each set of four cations with each anion, and simple quinary, senary, etc., systems by the combination of each set of five, six, etc., cations with each anion.

It is obvious that systems of the type $6//3$ cannot have simple systems containing more than 6 components.

Quaternary reciprocal systems are formed by the combination of each set of three cations with each pair of anions and of each set of three anions with each pair of cations. Starting from senary systems of the third class, however, it is necessary to distinguish, among the constituent reciprocal systems, systems of the second or third classes. Thus for the systems $4//3$, reciprocal systems of the second class are produced from each pair of anions with each set of four cations, and systems of the third class are produced from each set of three cations with all three anions.

The structure of systems $4//3$, $5//3$, $6//3$, or in general $C//3$, where C is any integer greater than 6, can be represented as shown in Table 5.

Since $C_3^1 = 3$, $C_3^2 = 3$, and $C_3^3 = 1$, it can readily be seen that in this case also there is a relationship between the table expressing the structure of the systems, and Pascal's triangle. This relationship, however, becomes increasingly complex on going to systems of higher classes.

Which multidimensional figures should be chosen for the representation of systems of the third class? It is known that the simplest representatives — quinary systems $3//3$ formed by nine simple salts — are represented by means of a four-dimensional nine-vertex figure — a prismatic hexahedroid — which is a section of a six-dimensional simplex, taken through the middle of its edges [6]. The structure of this figure can be represented as three triangles positioned in four-dimensional space relative to one another in the form of a triangle.

TABLE 5. The Structure of Systems of the Types 4//3, 5//3, 6//3, C//3

Lower constituent systems	Formulas for calculating the total number of systems			
	4//3	5//3	6//3	K//3
One-component	$C_4^1 \cdot C_3^1$	$C_5^1 \cdot C_3^1$	$C_6^1 \cdot C_3^1$	$C_C^1 \cdot C_3^1$
Binary	$C_4^2 \cdot C_3^1 + C_4^1 \cdot C_3^2$	$C_5^2 \cdot C_3^1 + C_5^1 \cdot C_3^2$	$C_6^2 \cdot C_3^1 + C_6^1 \cdot C_3^2$	$C_C^2 \cdot C_3^1 + C_C^1 \cdot C_3^2$
Ternary	$C_4^3 \cdot C_3^1 + C_4^2 \cdot C_3^2 + C_4^1 \cdot C_3^3$	$C_5^3 \cdot C_3^1 + C_5^2 \cdot C_3^2 + C_5^1 \cdot C_3^3$	$C_6^3 \cdot C_3^1 + C_6^2 \cdot C_3^2 + C_6^1 \cdot C_3^3$	$C_C^3 \cdot C_3^1 + C_C^2 \cdot C_3^2 + C_C^1 \cdot C_3^3$
Quaternary	$C_4^4 \cdot C_3^1 + C_4^3 \cdot C_3^2 + C_4^2 \cdot C_3^3$	$C_5^4 \cdot C_3^1 + C_5^3 \cdot C_3^2 + C_5^2 \cdot C_3^3$	$C_6^4 \cdot C_3^1 + C_6^3 \cdot C_3^2 + C_6^2 \cdot C_3^3$	$C_C^4 \cdot C_3^1 + C_C^3 \cdot C_3^2 + C_C^2 \cdot C_3^3$
Quinary	$C_4^4 \cdot C_3^2 + C_4^3 \cdot C_3^3$	$C_5^5 \cdot C_3^1 + C_5^4 \cdot C_3^2 + C_5^3 \cdot C_3^3$	$C_6^5 \cdot C_3^1 + C_6^4 \cdot C_3^2 + C_6^3 \cdot C_3^3$	$C_C^5 \cdot C_3^1 + C_C^4 \cdot C_3^2 + C_C^3 \cdot C_3^3$
Senary	$C_4^4 \cdot C_3^3$	$C_5^5 \cdot C_3^2 + C_5^4 \cdot C_3^3$	$C_6^6 \cdot C_3^1 + C_6^5 \cdot C_3^2 + C_6^4 \cdot C_3^3$	$C_C^6 \cdot C_3^1 + C_C^5 \cdot C_3^2 + C_C^4 \cdot C_3^3$
Septenary		$C_5^5 \cdot C_3^3$	$C_6^6 \cdot C_3^2 + C_6^5 \cdot C_3^3$	$C_C^7 \cdot C_3^1 + C_C^6 \cdot C_3^2 + C_C^5 \cdot C_3^3$
C-component				$C_C^C \cdot C_3^1 + C_C^{C-1} \cdot C_3^2 + C_C^{C-2} \cdot C_3^3$
(C+1)-component				$C_C^C \cdot C_3^2 + C_C^{C-1} \cdot C_3^3$

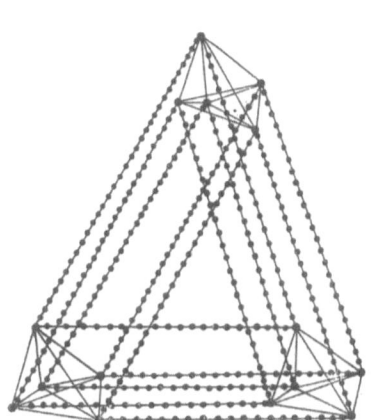

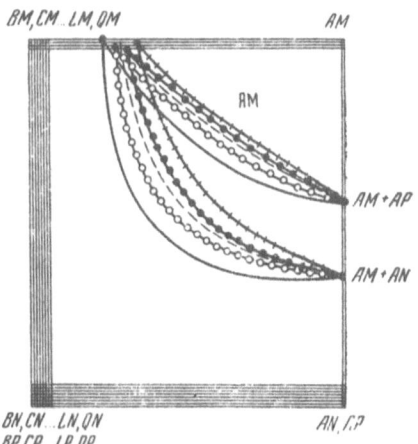

Fig. 17. Six-dimensional geometric figure for the representation of seven-component systems of the third class 5//3.

Fig. 18. Optimum projection of n-dimensional figures representing systems of the third class, on the plane of a diagram.

Similarly, for the representation of six-component systems 4//3 it is necessary to construct a five-dimensional figure (obtained as the section of a seven-dimensional simplex) consisting of three tetrahedra positioned in five-dimensional space in the form of a triangle. Systems C//3 can be represented by the multidimensional analog of a prismatic hexahedroid, in which three $(C-1)$-dimensional simplexes (i.e., simplexes with C vertices) are positioned relative to one another in $(C+1)$-dimensional space in the form of a triangle (Fig. 17).

It is evident that the structure of these figures corresponds completely to the structure of the systems which they are meant to represent. The Euler—Poincaré equation can therefore be used to confirm the accuracy of the formulas given in Table 5.

By carrying out the appropriate calculation of the total number of one-component, binary, ternary, etc., systems, we obtain the following results:

for systems 4//3
$$12 - 30 + 34 - 21 + 7 = 1 - (-1)^5; \ 2 = 2;$$

for systems 5//3
$$15 - 45 + 65 - 55 + 28 - 8 = 1 - (-1)^6; \ 0 = 0;$$

for systems 6//3
$$18 - 63 + 111 - 120 + 83 - 36 + 9 = 1 - (-1)^7; \ 2 = 2.$$

In all cases the formulas which we have derived correspond exactly to this equation, indicating that the former are correct.

Thus multidimensional figures suitable for the representation of systems of the third class with any number of components are formed as sections of the corresponding simplexes. This "breakdown" was first carried out by V. P. Radishchev [21], who showed that the simplest of these figures (a four-dimensional prismatic hexahedroid) can be broken down into six pentatopes by means of six sphenoids or half-pyramids in five different ways, depending on the thermochemical type to which the system belongs. The structure of systems of the third class becomes increasingly complex with increase in the number of components.

In V. P. Radishchev's papers, characteristic "dendrites"* are introduced. These indicate the relationship between the n-dimensional cells of the original n-dimensional figure and those $(n-1)$-dimensional figures by means of which the "breakdown" is carried out [4].

___
*In the literature they are called stable complexes or singular stars.

A characteristic feature is that the general form of these "dendrites" also becomes more complex as the class of the system becomes higher. For figures representing systems of the second class the "dendrites" give a linear relationship, for figures representing systems of the third class they include planar (two-dimensional) formations, for figures representing systems of the fourth class they include volume (three-dimensional) cells, etc.

The problem of the methods used to "break down" multidimensional figures of different types is of considerable importance in the study of multicomponent systems, since when no solid solutions are present each n-dimensional cell corresponds to an $(n+1)$-fold invariant point (for $p = \text{const}$). At the present time these methods are being successfully developed by N. S. Dombrovskaya for specific systems of higher classes.

The Optimum Projections of Figures Analogous to a Prismatic Hexahedroid

The optimum projections of multidimensional figures of this class are determined, as in the previous case, on the basis of the general relationships governing the formation of the optimum projections of the analogous four-dimensional figure. In the examination of the properties of the prismatic hexahedroid it was established that its optimum projection on the plane of a diagram belongs to the third type and is formed when the projection rays are parallel to one of its square faces. It is evident that with increase in the dimensionality of the original figure, the "parallel" unit should be one of its three-, four-, etc., dimensional cells. In the general case, if the figure represents a system C//3, this will be a cell whose dimensionality is smaller by two units. It must, however, contain square faces. A projection of this type is shown in Fig. 18. It can be seen that in this case the projection was carried out by means of rays parallel to a figure representing a lower constituent system of the second class, i.e., analogous to a tetrahedral hexahedroid. Figures of this type, however, must contain square faces, which are a distinguishing characteristic of the prisms. If the original figure representing the $(C + 2)$-component system C//3 was a $(c+1)$-dimensional figure, the figure forming part of the original figure and parallel to which the projection rays are drawn represents a system $(C-1)//2$, i.e., it is a $(c-1)$-dimensional figure.

The projection given in Fig. 18 shows all the characteristics of an optimum projection: 1) the coincident elements have undergone contraction to equal extents; 2) one of the vertices AM representing one of the original simple salts of the system is not merged with other vertices; and 3) the coincident elements are those elements of the figure corresponding to the ranges of crystallization of the same phase, formed by the salt AM (or of all the phases of the system containing this salt). As a result, the projection indicates the minimum and maximum boundaries of the ranges of crystallization of this phase and makes it possible to carry out all the quantitative calculations involved in the practical utilization of the phase diagrams (or composition—property diagrams).

The projection in Fig. 18 makes it possible to carry out quantitative calculations only within the limits of the ranges of crystallization of phases containing only one of the original simple salts of the system. Since a system C//3 contains a total of 3C simple salts, its complete representation would require the construction of at least 3C diagrams of the optimum type.

OPTIMUM PLANAR PROJECTIONS OF MULTIDIMENSIONAL FIGURES REPRESENTING SYSTEMS OF HIGHER CLASSES

The Structure of Systems of the Fourth Class [22]

The structure of systems of the fourth class is more complex than that of systems of the first, second, and third classes, examined above, because of the increase in the total number of double decomposition reactions. In order to derive appropriate formulas defining the number and nature of the lower constituent systems, we use the same arguments as in the previous cases. Thus it can readily be seen that the number of one- and two-component systems is formed as a result of the combination of each cation with each anion, or of each pair of cations with each anion and of each pair of anions with each cation, respectively.

Simple ternary, quaternary, etc., systems are formed by the combination of each set of three cations with each anion and of each set of three anions with each cation, or by the combination of each set of four cations with each anion and of all four anions with each cation, etc.

Starting from three-component systems, reciprocal systems may be formed in addition to simple systems: ternary reciprocal systems by the combination of each pair of cations with each pair of anions, and quaternary by the combination of three cations with two anions or of three anions with two cations.

In systems containing five or more components, reciprocal systems of two classes — second and third — are produced. In systems containing seven or more components, lower reciprocal systems of the fourth class are also produced.

The general relationship governing the structure of multicomponent systems of the fourth class is readily found by comparing the formulas defining the structure of systems 4//4, 5//4, and 6//4 (Table 6). Since $C_4^1 = 4$, $C_4^2 = 6$, $C_4^3 = 4$, and $C_4^4 = 1$, we can see the relationship between the formulas in Table 6 and Pascal's triangle. A fundamental difference, however, is that the structure of reciprocal systems of the fourth class is given by a sum whose terms are the numbers of Pascal's triangle, multiplied by 4, 6, 4, or 1, respectively. The representation of systems of the fourth class evidently requires geometric figures of a type different from those used to represent systems of the previous three classes.

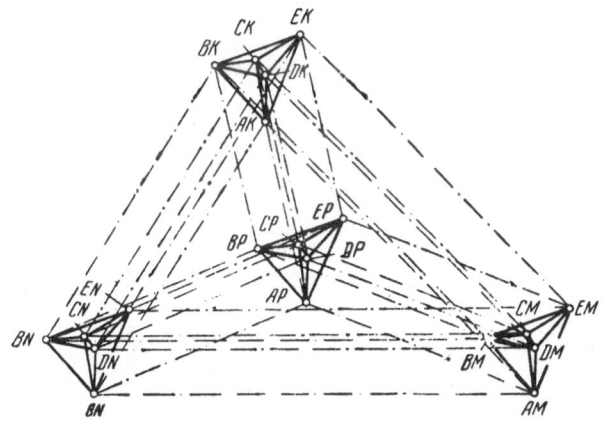

Fig. 19. Seven-dimensional figure for the representation of eight-component systems of the fourth class 5//4.

TABLE 6. The Structure of Systems of the Types 4//4, 5//4, 6//4, and C//4

Lower constituent systems	Formulas for calculating the total number of systems			
	4//4	5//4	6//4	K//4
One-component	$C_4^1 \cdot C_4^1$	$C_5^1 \cdot C_4^1$	$C_6^1 \cdot C_4^1$	$C_C^1 \cdot C_4^1$
Binary	$C_4^2 \cdot C_4^1 + C_4^1 \cdot C_4^2$	$C_5^2 \cdot C_4^1 + C_5^1 \cdot C_4^2$	$C_6^2 \cdot C_4^1 + C_6^1 \cdot C_4^2$	$C_C^2 \cdot C_4^1 + C_C^1 \cdot C_4^2$
Ternary	$C_4^3 \cdot C_4^1 + C_4^2 \cdot C_4^2 + C_4^1 \cdot C_4^3$	$C_5^3 \cdot C_4^1 + C_5^2 \cdot C_4^2 + C_5^1 \cdot C_4^3$	$C_6^3 \cdot C_4^1 + C_6^2 \cdot C_4^2 + C_6^1 \cdot C_4^3$	$C_C^3 \cdot C_4^1 + C_C^2 \cdot C_4^2 + C_C^1 \cdot C_4^3$
Quaternary	$C_4^4 \cdot C_4^1 + C_4^3 \cdot C_4^2 + C_4^2 \cdot C_4^3 + C_4^1 \cdot C_4^4$	$C_5^4 \cdot C_4^1 + C_5^3 \cdot C_4^2 + C_5^2 \cdot C_4^3 + C_5^1 \cdot C_4^4$	$C_6^4 \cdot C_4^1 + C_6^3 \cdot C_4^2 + C_6^2 \cdot C_4^3 + C_6^1 \cdot C_4^4$	$C_C^4 \cdot C_4^1 + C_C^3 \cdot C_4^2 + C_C^2 \cdot C_4^3 + C_C^1 \cdot C_4^4$
Senary	$C_4^4 \cdot C_4^2 + C_4^3 \cdot C_4^3 + C_4^2 \cdot C_4^4$	$C_5^5 \cdot C_4^1 + C_5^4 \cdot C_4^2 + C_5^3 \cdot C_4^3 + C_5^2 \cdot C_4^4$	$C_6^5 \cdot C_4^1 + C_6^4 \cdot C_4^2 + C_6^3 \cdot C_4^3 + C_6^2 \cdot C_4^4$	$C_C^5 \cdot C_4^1 + C_C^4 \cdot C_4^2 + C_C^3 \cdot C_4^3 + C_C^2 \cdot C_4^4$
Septenary	$C_4^4 \cdot C_4^3 + C_4^3 \cdot C_4^4$	$C_5^5 \cdot C_4^2 + C_5^4 \cdot C_4^3 + C_5^3 \cdot C_4^4$	$C_6^6 \cdot C_4^1 + C_6^5 \cdot C_4^2 + C_6^4 \cdot C_4^3 + C_6^3 \cdot C_4^4$	$C_C^6 \cdot C_4^1 + C_C^5 \cdot C_4^2 + C_C^4 \cdot C_4^3 + C_C^3 \cdot C_4^4$
C-component*				$C_C^C \cdot C_4^1 + C_C^{C-1} \cdot C_4^2 + C_C^{C-2} \cdot C_4^3 + C_C^{C-3} \cdot C_4^4$
(C+1)-component*				$C_C^C \cdot C_4^2 + C_C^{C-1} \cdot C_4^3 + C_C^{C-2} \cdot C_4^4$
(C+2)-component*				$C_C^C \cdot C_4^3 + C_C^{C-1} \cdot C_4^4$

*C is any integer greater than six.

The simplest form of reciprocal system of the fourth class — the systems 4//4 — contain seven components and hence can be represented by six-dimensional figures. They therefore have no analogs among four-dimensional geometric figures. At the same time the study of multidimensional figures representing systems of the first, second, and third classes indicate the nature of the figure by means of which it is possible to represent the systems 4//4. This should be a unique six-dimensional tetrahedron, each vertex of which is in turn a tetrahedron. The general tetrahedral disposition of the complex vertices is characteristic of all figures representing systems of the fourth class, just as the general triangular disposition is characteristic of figures representing systems of the third class. In both cases the actual vertices are simplexes for which the number of vertices is equal to the number of cations in the system.

Figure 19 shows a seven-dimensional figure suitable for the representation of eight-component systems of the fourth class (i.e., of the type 5//4). Since the structure of the systems and the structure of the figures representing them should be identical, the Euler—Poincaré formula can be used to verify the accuracy of the formulas given in Table 6.

We obtain the following values:

for systems of the type 4//4

$$16 - 48 + 68 - 56 + 28 - 8 = 1 - (-1)^6; \ 0 = 0;$$

for systems of the type 5//4

$$20 - 70 + 120 - 125 + 84 - 36 + 9 = 1 - (-1)^7; \ 2 = 2;$$

for systems of the type 6//4

$$24 - 96 + 194 - 246 + 209 - 120 + 45 - 10 = 1 - (-1)^8; \ 0 = 0.$$

In all cases agreement is complete, so that the structure of systems of the fourth class has been correctly determined.

The Structure of Systems of Any Class [23]

The examination of the methods of formation of lower constituent systems in multicomponent systems of the first four classes is applicable in principle to systems of any class and can be generalized in the following expressions. Let us take a system C//A, where C, the number of cations, and A, the number of anions, are any integers greater than 4. The class of this system is evidently determined by the smaller of these two numbers, and its total number of components is determined by the sum C + A, minus one, since the sum of the cations is equal to the sum of the anions of any mixture of salts. The number of one-component, binary, ternary, quaternary, etc., constituent systems is determined from the number of combinations possible between the cations and anions of the system taken singly, in pairs, in threes, etc. Reciprocal systems are subdivided into an increasing number of classes with increase in C and A, i.e., as the class of the system becomes higher. In the general case, in the system C//A we may have reciprocal systems of the type X//Y, where X ≤ C and Y ≤ A.

Let us take as an example the system ABCD//XYZ. For this system, reciprocal systems of the second and third classes are evidently possible. Reciprocal systems of the second class are formed, firstly, by the combination of each pair of cations with each pair of anions. These are the ternary systems: AB//XY, AC//XY, AD//XY, BC//XY, CD//XY, BD//XY, AB//XZ, AC//XZ, AD//XZ, BC//XZ, BD//XZ, CD//XZ, AB//YZ, AC//YZ, AD//YZ, BC//YZ, BD//YZ, and CD//YZ. Secondly, they are formed by the combination of each pair of cations with all three anions and of each pair of anions with each set of three cations. These are the quaternary systems: AB//XYZ, AC//XYZ, AD//XYZ, BC//XYZ, BD//XYZ, CD//XYZ, ABC//XY, ABC//YZ, ABC//XZ, ABD//XY, ABD//XZ, ABD//YZ, BCD//XY, BCD//XZ, BCD//YZ, ACD//XY, ACD//XZ, and ACD//YZ.

Reciprocal systems of the third class are formed by the combination of each set of three cations with all three anions. These are the quinary systems: ABC//XYZ, ABD//XYZ, ACD//XYZ, and BCD//XYZ.

TABLE 7. The Structure of Multicomponent Reciprocal Systems C//A of Any Class

Lower constituent systems	Formulas for calculating the total number of systems
One-component	$C_C^1 \cdot C_A^1$
Binary	$C_C^2 \cdot C_A^1 + C_C^1 \cdot C_A^2$
Ternary	$C_C^3 \cdot C_A^1 + C_C^2 \cdot C_A^2 + C_C^1 \cdot C_A^3$
Quaternary	$C_C^4 \cdot C_A^1 + C_C^3 \cdot C_A^2 + C_C^2 \cdot C_A^3 + C_C^1 \cdot C_A^4$
Quinary	$C_C^5 \cdot C_A^1 + C_C^4 \cdot C_A^2 + C_C^3 \cdot C_A^3 + C_C^2 \cdot C_A^4 + C_C^1 \cdot C_A^5$
Senary	$C_C^6 \cdot C_A^1 + C_C^5 \cdot C_A^2 + C_C^4 \cdot C_A^3 + C_C^3 \cdot C_A^4 + C_C^2 \cdot C_A^5 + C_C^1 \cdot C_A^6$
m-component*	$C_C^m \cdot C_A^1 + C_C^{m-1} \cdot C_A^2 + C_C^{m-2} \cdot C_A^3 + \ldots + C_C^{m-m/2+1} \cdot C_A^{m/2} + C_C^{m/2} \cdot C_A^{m-m/2+1} +$ $+ \ldots + C_C^3 \cdot C_A^{m-2} + C_C^2 \cdot C_A^{m-1} + C_C^1 \cdot C_A^m$
(m+1)-component	$C_C^{m+1} \cdot C_A^1 + C_C^m \cdot C_A^2 + \ldots + C_C^{m-m/2+1} \cdot C_A^{m/2+1} + \ldots + C_C^2 \cdot C_A^m$
p-component†	$C_C^p \cdot C_A^1 + C_C^{p-1} \cdot C_A^2 + \ldots + C_C^{p-p/2+1} \cdot C_A^{p/2} + C_C^{p/2} \cdot C_A^{p-p/2+1}$
(p+1)-component	$C_C^p \cdot C_A^2 + C_C^{p-1} \cdot C_A^3 + \ldots + C_C^{p-p/2+1} \cdot C_A^{p/2+1}$
(C+A−2)-component	$C_C^C \cdot C_A^{A-1} + C_C^{C-1} \cdot C_A^A$

* $m \leq A < C$ and an even number.

† $p \leq C > A$ and an even number

Note. If m is an odd number, the number of m-component systems is calculated from the formulas:

$$C_C^m \cdot C_A^1 + C_C^{m-1} \cdot C_A^2 + \ldots\; C_C^{m-(m-1)/2} \cdot C_A^{(m-1)/2} + C_C^{(m-1)/2+1} \cdot C_A^{(m-1)/2+1} + \ldots C_C^2 \cdot C_A^{m-1} + C_C^1 \cdot C_A^m.$$

The calculation of the number of (m+1)-component systems is also changed and is carried out in this case from the formulas:

$$C_C^{m+1} \cdot C_A^1 + C_C^m \cdot C_A^2 + \ldots C_C^{m-(m-1)/2+1} \cdot C_A^{(m-1)/2+1} + \ldots + C_C^2 \cdot C_A^m.$$

The formulas for the calculation of the lower constituent systems of p and (p+1) components is changed in analogous fashion if p is an odd number.

The original system ABCD//XYZ is obviously a system of the third class. Its lower constituent systems evidently cannot contain systems of the fourth, fifth, or higher classes.

We shall not deal with the method for determining the number of one-component, binary, ternary, quaternary, quinary, and senary constituent systems, since this is more or less obvious, but shall deal merely with some general cases. Thus the number of simple m-component systems for $m \leq C$ (if $C < A$) or for $m \leq A$ (if $A < C$) is always equal to $C_C^m \cdot C_A^1 + C_A^m \cdot C_C^1$. Let us assume that $A < C$. In this case, starting from the $(m+1)$-component systems, the number of simple salts is determined entirely by the expression $C_C^{m+1} \cdot C_A^1$, since the second expression $C_A^{m+1} \cdot C_C^1$ becomes meaningless, just as, for example, the number of combinations of three elements, taken in fours, is meaningless. Similarly, all the expressions characterizing the number of reciprocal systems with two, three, four, etc., ions of the same sign, in which one of the factors is C_A^{m+1}, for example, $C_A^{m+1} \cdot C_C^2$ or $C_A^{m+1} \cdot C_C^3$, becomes meaningless. Thus $(m+2)$-component systems cannot contain systems whose number is determined by the expression $C_A^{m+1} \cdot C_C^2$, and in $(m+3)$-component systems, all systems characterized by the term $C_A^{m+1} \cdot C_C^3$ disappear, etc. The same takes place with further increase in the number of components of the lower constituent systems. First, all $(C+1)$-component simple systems disappear, since the number C_C^{C+1} characterizing them becomes meaningless, after which we have the gradual disappearance of all reciprocal systems with two, three, etc., ions of the same sign, whose numbers are determined by means of formulas containing similar expressions as one of the factors.

These principles may be illustrated for the case of the system ABCD//XYZ. Here $A = 3$ and $C = 4$, i.e., $A < C$, so that $m \leq A$ represents a number less than or equal to three. It is evident that the expression C_A^m is possible only for $m =$ one, two, or three; for $m =$ four, the expression becomes meaningless, and the expression is also meaningless for $m > 4$. Thus the number of simple quaternary systems of the system ABCD//XYZ is determined entirely by the expression $C_C^4 \cdot C_A^1 = C_4^4 \cdot C_3^1$, since $C_A^4 \cdot C_C^1 = C_3^4 \cdot C_4^1$ contains the meaningless factor C_3^4. The number of reciprocal quinary systems with two ions of the same sign (of the second class) is determined in this case entirely by the expression $C_C^4 \cdot C_A^2 = C_4^4 \cdot C_3^2$, since the second term $C_A^4 \cdot C_C^2 = C_3^4 \cdot C_4^2$ contains the same expression C_3^4, with no physical significance. Finally, the number of senary systems with three ions of the same type (of the third class) is determined by the expression $C_C^4 \cdot C_A^3 = C_4^4 \cdot C_3^3$; the second term in the general formula $C_A^4 \cdot C_C^3 = C_3^4 \cdot C_4^3$ contains the same factor C_3^4, equal to zero.

The expressions determining the number of lower constituent systems in multicomponent systems of any class are compared in Table 7.

The formulas given indicate not only the total number of constituent systems with a given number of components but also the number of systems of each class. For this purpose it is sufficient to combine all the terms for which the number of combinations of one, two, three, etc., is taken from C or A, since they define, respectively, systems of the first, second, third, or in general m-th class.

Let us carry out the calculation of the total number of 17-component systems in the 36-component systems 20//17 using the formulas in Table 7. Here we can combine the expressions for systems of the same class, and since $C = 20$ and $A = 17$ (and $m = 17 = A$ is an odd number), we use the expression given in the footnote to Table 7:

$$(C_{20}^{17} \cdot C_{17}^1 + C_{20}^1 \cdot C_{17}^{17}) + (C_{20}^{16} \cdot C_{17}^2 + C_{20}^2 \cdot C_{17}^{16}) + (C_{20}^{15} \cdot C_{17}^3 + C_{20}^3 \cdot C_{17}^{15}) + \ldots + (C_{20}^{17-16/2} C_{17}^{16/2+1}).$$

Here the first two terms refer to simple systems of the first class and all the subsequent terms to reciprocal systems of different classes from the second to the ninth inclusive. The total number of 20-component systems of the same system 20//17 is calculated in analogous fashion from the formula given in Table 7 for the calculation of p-component systems if $p \leq C$ and is an even number. This is equal to

$$(C_{20}^{20} \cdot C_{17}^1) + (C_{20}^{19} \cdot C_{17}^2) + \ldots + (C_{20}^{20-10+1} \cdot C_{17}^{10} + C_{20}^{10} \cdot C_{17}^{20-10+1}).$$

Here also the first term characterizes the total number of simple 20-component systems, and all the subsequent terms characterize the numbers of reciprocal systems of different classes from the second to the

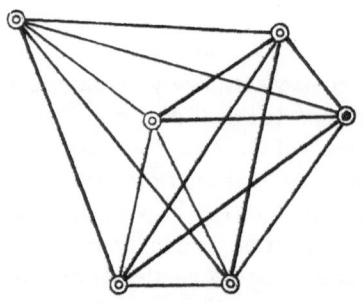

Fig. 20. Multidimensional geometric figure for the representation of multi-component reciprocal systems of the sixth class C//6.

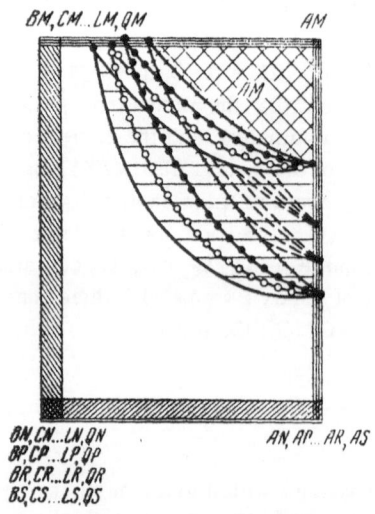

Fig. 21. Optimum projection of n-dimensional figures representing multi-component systems of any class, on the plane of a diagram.

tenth. Finally, the total number of 35-component systems "bounding" the 36-component system under consideration is calculated from the last formula in Table 7. It is equal to

$$C_{20}^{20} \cdot C_{17}^{16} + C_{20}^{19} \cdot C_{17}^{17} = 37.$$

Thus a system of m components of a higher class is always bounded by $(m-1)$-component reciprocal systems, the number of which is equal to the sum of the numbers of cations and anions forming the given system.

As the class of the systems becomes higher, their structure becomes exceptionally complex. The following data provide sufficient indication.

In a 36-component simple system there is a total of 1,947,792 six-component systems. In a system 20//17 with the same total number of components, however, there are 10,198,504 six-component systems, including 906,440 simple systems, 3,284,264 reciprocal systems of the second class, and 6,007,800 reciprocal systems of the third class.

The Optimum Projections of Multidimensional Figures Corresponding to Systems of Higher Classes

If the number of ions of the same sign in multicomponent reciprocal systems is greater than 3, i.e., for systems of the fourth, fifth, and higher classes, there is no suitable four-dimensional figure which might serve as a prototype for analogous figures of higher dimension. For the case of systems of the fourth class, however, it can be seen that this is no obstacle, and that it is possible to select figures suitable in each specific case for the representation of the compositions of different systems and the construction of their phase diagrams. In the case of systems of the fourth class, these figures were found on the basis of an analysis of the results obtained in the selection of the corresponding figures for systems of the first, second, and third classes.

The multidimensional figures found, and the ways in which they are formed, serve in turn as a basis for dealing with systems of any class with any number of components. If we use for the representation of systems 4//4 a figure consisting of four tetrahedra which themselves form a tetrahedron in six-dimensional space, it is possible for systems 5//5 to imagine an analogous eight-dimensional figure consisting of five pentatopes forming a pentatope, and for systems 6//6, similarly, six hexatopes forming a hexatope in ten-dimensional space, and so on to infinity.

At the same time the general principles of the construction of these geometric figures become clear. In the general case, the representation of a system C//A requires a multidimensional $(C + A - 2)$-dimensional figure consisting of a combination of simplexes in the form of a simplex in multidimensional space. This is a multidimensional simplex for which the number of vertices is A, but each vertex is not simply a point but a characteristic simplex for which the number of vertices is C. Figure 20 gives a schematic representation of a figure of this type, representing a multicomponent system of the sixth class C//6.

Which are the optimum projections of these figures? It is found that they can be predicted on the basis of the general principles governing the formation of the optimum projections of the geometric figures which were examined above. Of particular importance are the results of the study of the way in which the optimum projection of the prismatic hexahedroid is formed. This figure, as noted above, served as a general model for all other multidimensional figures suitable for the representation of reciprocal systems of higher classes. In analogous fashion, the optimum projection of the four-dimensional prismatic hexahedroid can be made the basis of the construction of the optimum projections of all these multidimensional figures. It was found (see page 27) that for the construction of the optimum planar projections of figures representing systems of the third class it is necessary that the projection rays be parallel to one of the constituent figures, with dimensionality lower by 2 than the dimensionality of the original figure. Moreover, it is necessary that this "cell" parallel to the rays should contain square faces (see Fig. 18). Essentially, we have here formulated the conditions under which we obtain the optimum planar projection of a multidimensional figure representing any system of higher class. Figure 21 shows the projection obtained in this way for a figure representing the system ABC ... LG//MNP ... RS [24]. It can readily be seen that it is in fact an optimum projection since all the co-incident elements in it have undergone contraction to equal extents, and one of the vertices — AM — has not merged with any other vertex of the original figure. All this makes it possible to represent the components of the system on the same scale and to represent the boundaries of the range of crystallization of one of the phases of the system being studied. Since the minimum number of phases of the system (in the absence of solid solutions) is equal to the number of its simple salts, i.e., in the general case $C \cdot A$, the representation of the system as a whole requires the construction of at least $C \cdot A$ diagrams of the optimum-projection type.

OPTIMUM PROJECTIONS OF SOME FOUR-DIMENSIONAL FIGURES IN THREE-DIMENSIONAL SPACE [25, 26]

It is sometimes useful to use three-dimensional models for the representation of multicomponent systems.

If it is possible in such a model to represent any three-dimensional section of the system, or if some of the components, for any reasons, can be omitted, it does not differ in principle from the usual models of quaternary systems.

If, however, it is necessary to represent a multicomponent system as a whole, it is important, as in the case of planar diagrams, to select the optimum projection of the corresponding four-dimensional figure into coordinate space.

Let us examine this in more detail for the case of four-dimensional figures representing quinary systems.

Every four-dimensional figure has a total of four three-dimensional projections.* These are constructed on the basis of the values of the coordinates of the vertices, which are calculated by definite methods [6]. These values are plotted on coordinate axes forming different coordinate spaces, and the figurative points obtained for each vertex are connected by means of straight lines in three-dimensional space in exactly the same order as that in which the vertices of the four-dimensional figure are connected by the edges.

Optimum Model of a Pentatope

Of the four projections of a pentatope into coordinate spaces, three (Figs. 22a, b, and d) are projections of the first type since they are obtained by projection by means of rays which are not parallel to any of its edges. In all the figures the model is a tetrahedron whose vertices are formed by four of the vertices of the pentatope, its fifth vertex being situated either in the internal volume of the tetrahedron (see Figs. 22a and b) or on one of its edges (see Fig. 22d). In all three cases we have superposition and coincidence of parts of the figure, which make these projections unsuitable for the construction of the phase diagrams of chemical systems. The pentatope, however, gives another projection into coordinate space (see Fig. 22c), which is a projection of the second type, since it is obtained by projection by means of rays parallel to one of the edges of the original figure. The model is a tetrahedron, three of whose vertices are formed by three of the vertices of the pentatope (E, D, and C), the fourth vertex corresponding to the two remaining vertices of the original figure (A and B), merged into one. Thus the edge AB of the pentatope, parallel to the projection rays, degenerates into a point. Moreover, the following faces are completely superimposed in pairs in the model: ECA and ECB, DEA and DEB, and DCA and DCB.

Finally, three faces of the pentatope — ABE, ABD, and ABC — degenerate into straight lines. Here all the faces on the model undergo contraction to different extents, but since the completely coincident edges and faces of the original figure naturally undergo contraction to equal extents, this makes it possible to compensate the observed contraction by increasing the scale, and to construct a model in the form of a regular tetrahedron. Thus by using this projection for the construction of a model of a quinary system, it is possible to plot the concentrations of all its components on the same scale. On the other hand, although nine faces of the original pentatope are not reflected completely on the model, the tenth and last face DCE is represented individually and is not screened by other elements of the figure.

---

*Since the number of combinations of three from the total number of coordinate axes in this case is equal to $C_4^3 = 4$.

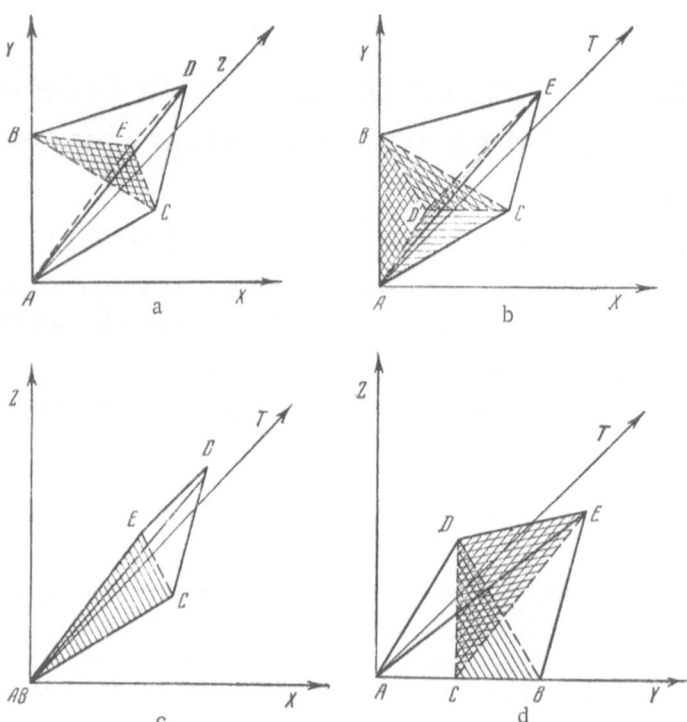

Fig. 22.   Projections of a pentatope into the coordinate spaces:
a) XYZ; b) XYT; c) XZT; d) YZT.

Since the faces of the pentatope represent the ternary systems forming part of the original quinary system, the individual components D, C, E, and also the three binary systems and the ternary system formed by these are represented completely on the model.

It is also evident that the regions of crystallization of the phases including these components can be represented in the internal volume of the tetrahedron adjoining the face EDC. The model can be represented as the combination of two adjacent tetrahedra of the pentatope, representing two quaternary systems — AEDC and BEDC — with a common ternary system EDC. The boundaries of the regions of crystallization of the phases rich in components E, D, and C (in each separately, in two of them, or in all three) will naturally be different in each of the above quaternary systems, and also in the quinary system as a whole. They all, however, adjoin the face EDC, and the model will give their most probable maximum and minimum limits.

Thus the projection in Fig. 22c is the optimum model of the pentatope. It makes it possible to represent the phase diagrams of quinary systems of the first class and to give a quantitative representation of the boundaries of the ranges of crystallization of the individual phases of the system. Since the model reflects completely three of the components, representation of the quinary system as a whole requires the construction of two models of the optimum type.

The Optimum Model of a Tetrahedral Hexahedroid

The projections into coordinate spaces of the tetrahedral hexahedroid, which can be used to represent quinary systems of the second class, for example ABCD//EF, include two projections of the first type and two of the second type.

The projections of the first type are shown in Figs. 23b and c. Both models are triangular prisms whose bases are formed by six vertices of the original figure. The remaining two vertices are situated separately either in the centers of the triangular bases (Fig. 23b) or on one of their edges (Fig. 23c). If we try to use models of this type to represent a corresponding system, we find that two of its eight salts (for example the

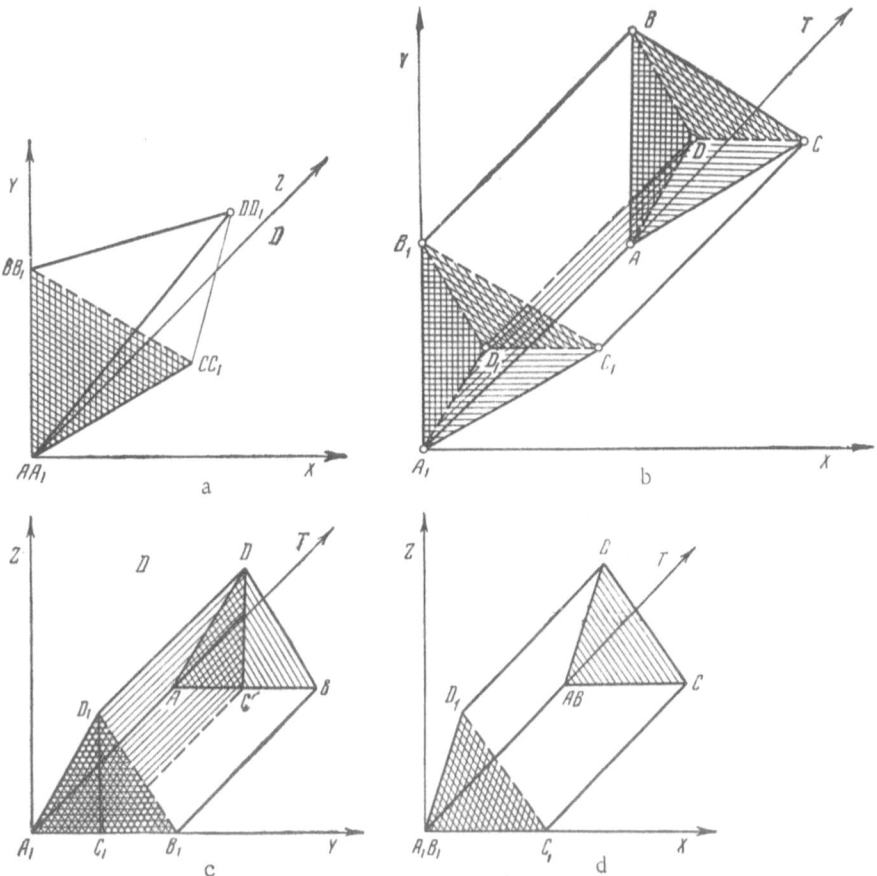

Fig. 23. Projections of a tetrahedral hexahedroid into the coordinate spaces:
a) XYZ; b) XYT; c) YZT; d) XZT.

salts DE and DF) would have to be plotted on a different scale from the other six. Moreover, the ranges of crystallization of all the phases containing these two salts would coincide completely with the ranges of crystallization of the other phases of the system. Thus these two projections of the first type are clearly unsuitable for the construction of diagrams satisfying the requirements of physicochemical analysis.

The tetrahedral hexahedroid, however, has another two projections, of the second type. One of these (Fig. 23a) is a tetrahedron, obtained by projection by means of rays parallel to four mutually parallel edges of the original figure — AA_1, BB_1, CC_1, and DD_1 (which form part of its square faces). As a result, all the vertices of the hexahedroid are merged in pairs on the model, since these edges are degenerate. The practical application of this model for the construction of the phase diagrams of quinary systems of the type 4//2 is extremely limited, since all eight simple salts of the system are represented jointly.

If we have a system ABCD//EF, such a model can give a picture of the quantitative relationships only in those cases where it is possible to restrict ourselves to data concerning the cations, independently of the anions.

The other three-dimensional projection of the tetrahedral hexahedroid, of the second type, is shown in Fig. 23d. This is a triangular prism, whose triangular bases are formed by all eight vertices of the original figure; four vertices are each represented individually — two each in the upper and lower bases — and the other four are combined in pairs and represented in pairs at each of the vertices of both bases. The model obtained is an optimum projection of the hexahedroid into three-dimensional space. It is obviously obtained by projection of the original figure by means of rays parallel to one of the edges forming part of the triangular faces of the hexahedroid.

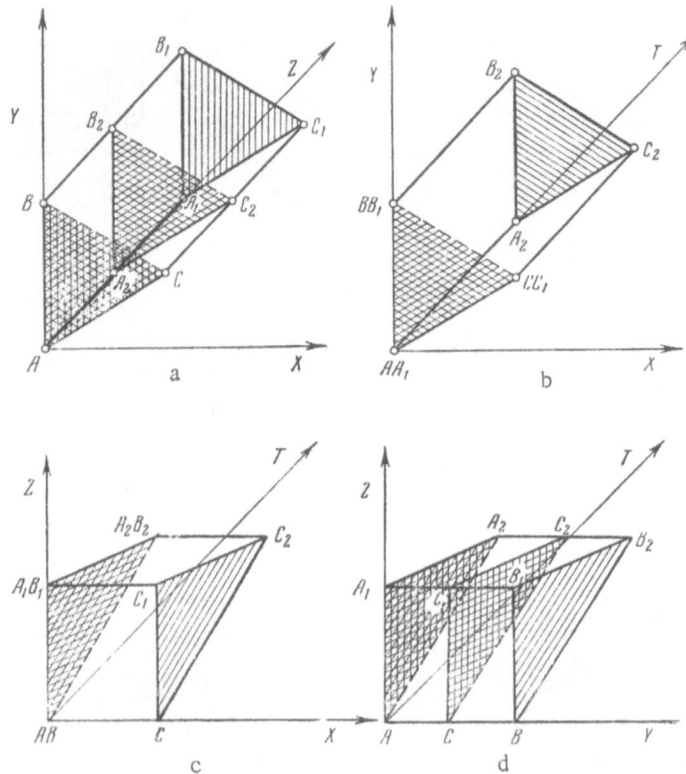

Fig. 24. Projections of a prismatic hexahedroid into the coordinate spaces:
a) XYZ; b) XYT; c) XZT; d) YZT.

If this model is used to represent the system ABCD//EF, four of its eight simple salts, for example, AE, BE, AF, and BF, can be examined. Thus the representation of the system as a whole requires the construction of two models of the optimum type.

The Optimum Model of a Prismatic Hexahedroid

The prismatic hexahedroid also gives two projections of the first type, which are triangular prisms (Figs. 24a and d). The upper and lower triangular bases of these prisms are formed by six of the vertices of the original figure — three vertices for each base. The other three vertices of the hexahedroid are situated in both prisms at the centers of its lateral edges. These two projections are therefore completely identical. They are unsuitable for the construction of diagrams for chemical systems, for the same reasons as in the previous analogous cases.

The prismatic hexahedroid, however, has two other projections, of the second type (Figs. 24b and c). These are identical and show no fundamental differences, since each can be obtained from the other by appropriate rearrangement of the components at the vertices of the figure. We shall therefore restrict ourselves to the examination of one of them, for example, Fig. 24b. This is obtained by projection of the prismatic hexahedroid into three-dimensional space by means of rays parallel to any three mutually parallel edges forming part of the triangular faces of the original figure. The model is a triangular prism, one of whose bases is formed by three vertices of the original figure, for example, A_2, B_2, and C_2, while the second base is formed by the six remaining vertices, combined in pairs — AA_1, BB_1, and CC_1. The equal contraction of the superimposed elements and the possibility of examining the parts of the figure adjoining three of its vertices, each of which is represented individually, makes this model an optimum model. If this model is used to represent the quinary system ABC//MNP, we can represent on the model the boundaries of the ranges of crystallization of all phases formed by any three simple salts of the system with a common cation or a common anion, for example, AM, BM, and CM. Thus the representation of the system as a whole requires the construction of two additional models of the optimum type: one of these gives the ranges of crystallization of phases containing the salts AN, BN, and CN, and the other gives the ranges of crystallization of phases containing the salts AP, BP, and CP.

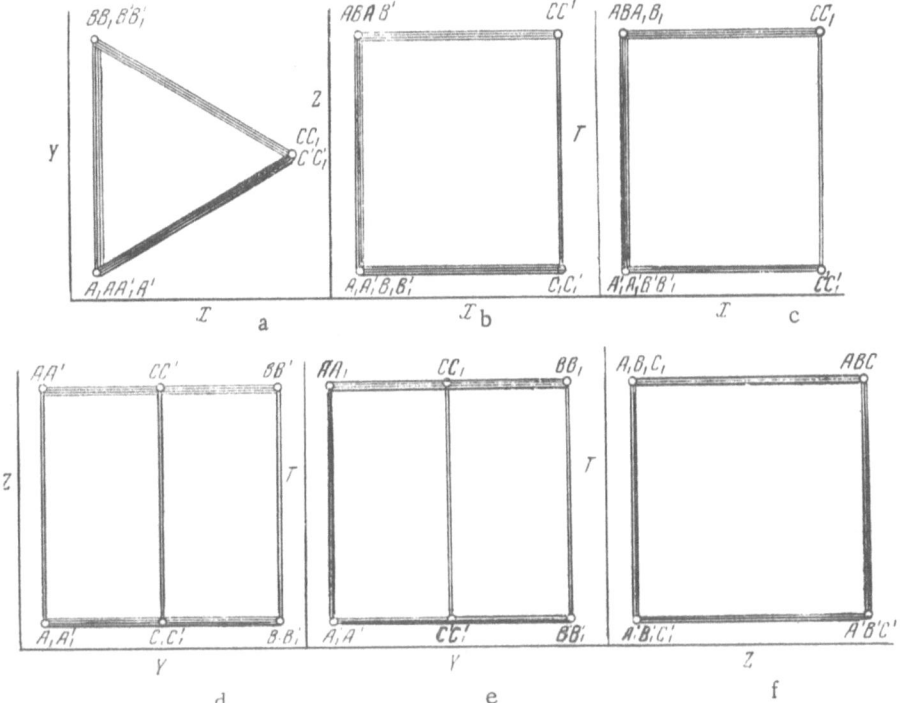

Fig. 25. Projections of a prismatic heptahedroid on six coordinate planes.

The Optimum Model of a Prismatic Heptahedroid

In the study of specific systems, in addition to the three four-dimensional figures — the pentatope, tetrahedral hexahedroid, and prismatic hexahedroid — which are the most suitable for the representation of five-component systems of the first, second, and third class, one other figure — the prismatic heptahedroid — is of considerable importance. It becomes necessary to use it in all cases where it is desired to represent a quinary system for which the independent variables include not only the concentrations of the components but also any other equilibrium factor (for example, temperature, pressure, or time) or property of the system.

It was found [6] that the prismatic heptahedroid has no optimum projections into coordinate space (Fig. 25). It can be seen from Fig. 25 that the six planar projections of the heptahedroid do not include projections of the first type. The two projections of the second type (Figs. 25d and e) are identical; they are obtained by projection of the original figure by means of rays parallel to edges forming part of its square faces. In both cases, all twelve vertices of the heptahedroid are superimposed in pairs on the projections; moreover, the individual superimposed edges undergo contraction to different extents as a result of this projection. The other four planar projections of the heptahedroid are projections of the third type. Two of these are obtained by projection of the heptahedroid by means of rays parallel to its square (Fig. 25a) or triangular (Fig. 25f) faces.

As a result, all the vertices of the original four-dimensional figure merge on these projections (in fours or in threes at each vertex of the projection) and this makes necessary the joint representation of the individual salts or other variables determining the state of the system.

Finally, two projections (Figs. 25b and c) are identical. They are obtained by projection of the heptahedroid by means of rays parallel to any two square faces of the heptahedroid. Since these two faces are also parallel to a further two edges of the heptahedroid, bounding its other faces, the degenerate elements on the resulting planar projection include not only the two "parallel" faces but also these two edges, and as a result none of the vertices of the original figure is represented individually. Thus the prismatic heptahedroid has no planar projections of the optimum type, suitable for quantitative calculations. Its projections into three-dimensional space, however, may include models suitable for the construction of the phase diagrams of corresponding systems.

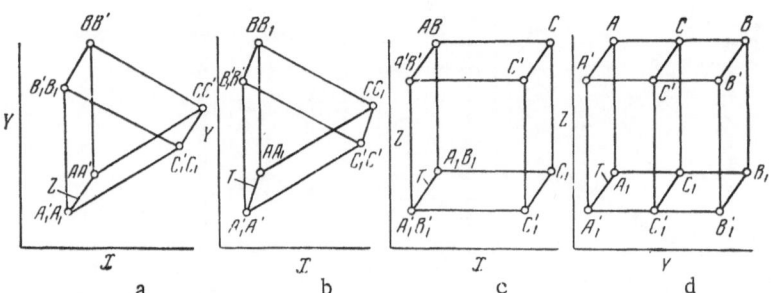

Fig. 26. Projections of a prismatic heptahedroid into the coordinate spaces:
a) XYZ; b) XYT; c) XZT; d) YZT.

Examinations of the projections of a prismatic heptahedroid into coordinate spaces (Fig. 26) reveals significant differences between them.

Here there are no projections of the first type, if we exclude the projection shown in Fig. 26d. In this model, each vertex of the heptahedroid is represented individually, so that all the edges of the original figure are represented on it. Its triangular faces, however, degenerate into straight lines. The unequal contraction of the edges and the superposition of parts corresponding to the ranges of crystallization of different phases of the system make the projection in Fig. 26d unsuitable in practice.

Projections of the second type are shown in Figs. 26a, b, and c. Of these, the first two are identical. They are triangular prisms obtained by projection of the heptahedroid by means of rays parallel to edges forming part of its square faces (horizontal [b] or vertical [a]). Since these horizontal (or vertical) edges are parallel, all the apices of the original figure are superimposed in pairs in both projections. Thus none of the independent variables is represented individually on either of these models; each is represented jointly, in combination with another variable. At the same time, independent variables of quite different character may be combined in pairs, for example, components with other equilibrium factors or with some properties.

The projection shown in Fig. 26c does not suffer from these disadvantages. This model is a cube obtained by projection of the original four-dimensional figure by means of rays parallel to edges forming part of its triangular faces. Although each of these edges is parallel to another three, so that with this method of projection four edges of the heptahedroid degenerate into points, i.e., eight of its vertices are represented in pairs, the remaining four vertices are each represented individually. Moreover, the superimposed faces undergo contraction to equal extents, so that overall this projection (Fig. 26c) is an optimum model of the prismatic heptahedroid.

The representation of the corresponding system as a whole obviously requires three models of the optimum type.

The Optimum Models for the Representation of Multicomponent Systems

The transition from the optimum models representing quinary systems to models suitable for the representation of systems with any number of components presents no difficulties.

The basic principles are the same as those of the corresponding transition for projections on the plane of a diagram. It is evident that for systems of the first, second, and third classes the models should be analogous to those described above for the pentatope, tetrahedral hexahedroid, and prismatic hexahedroid, the only difference being that in each case a correspondingly greater number of components or original salts must be represented jointly. It is necessary to represent individually: 1) any three components at the vertices of a tetrahedron — for systems of the first class; 2) any four original simple salts in twos at two of the vertices of a triangular prism — for systems of the second class; and 3) any three original simple salts at the vertices of one of the bases of the triangular prism — for systems of the third and subsequent classes.

PART II

EXAMPLES OF THE PREDICTION OF THE PROPERTIES OF MULTICOMPONENT SYSTEMS ON THE BASIS OF THE OPTIMUM PROJECTIONS METHOD

INTERPOLATION AND EXTRAPOLATION WITHIN THE RANGES
OF CRYSTALLIZATION OF IDENTICAL PHASES OF THE SYSTEM

As pointed out by N. S. Kurnakov, geometric methods have the important advantage that they make it possible to express the qualitative and quantitative relationship between properties and composition even in the case where the algebraic expression for this function is not known. At the same time it is known that the properties change continuously within the limits of the range of existence of a given phase and undergo sharp changes (inflections and breaks in the property curves) on changing from one phase to another. Thus if it is possible to determine the range of crystallization of a given phase, the determination of the general relationship governing the change in properties of this entire range as a whole is possible on the basis of a comparatively small number of reference points. Thus it is sufficient to make an experimental study of the properties of 3-4 compositions of the system, corresponding to these boundaries, in order to predict more or less accurately the properties of the system throughout the whole of this range.

These principles form the basis for the extensive practical application of the phase diagrams and composition—property diagrams of ternary and quaternary systems — metallic and salt, aqueous and nonaqueous, in solution and in melts.

On going to systems with more than four components, however, geometric difficulties were previously encountered in the application of diagrams. At the same time the experimental study of these multicomponent systems is exceptionally tedious, so that geometric methods are particularly advantageous in this case.

The method of optimum projections has advantages from this viewpoint. The graphs and models constructed by this method are comparatively simple and are in many respects similar to the graphs and models usually employed for systems with 3-4 components. A particularly important feature, however, is that they make it possible to predict the properties of multicomponent systems on the basis of data for the binary and ternary constituent systems. This prediction is based on the general character of the optimum projections of the geometric figures used for the representation of systems of any class with any number of components, however great.

The Optimum Projections of Three-Dimensional Geometric Figures

The optimum projection of the regular tetrahedron representing simple quaternary systems (Fig. 11b) can be obtained by the superposition and coincidence of two adjoining faces corresponding to two of the ternary systems making up the quaternary system; the binary eutectics corresponding to the points O, P, N, L, M, and K lie on the six edges of the tetrahedron, and the ternary eutectics corresponding to the points X, Y, Z, and U lie on its four faces (Fig. 27). The quaternary eutectic R evidently lies in the internal volume of the tetrahedron, which is thus divided into four volumes — the ranges of crystallization of the four solid phases of the system, formed by its components A, B, C, and D. They are bounded by the eutectic lines and surfaces defining, respectively, the ranges of the joint crystallization of two or three boundary phases. These eutectic lines and surfaces are usually curved to a greater or lesser extent, depending on the properties of the components and the binary and ternary systems which they form. By means of the optimum projection of the tetrahedron it is possible to show the boundaries of the range of crystallization of any of the solid phases of the system. Let us assume that we wish to represent the range of crystallization of the phase C adjoining the vertex C of the tetrahedron. It can be seen from Fig. 27 that it is bounded by the following plane surfaces forming part of the external boundary of the tetrahedron — NXLC, MULC, and MZNC — and also by curved surfaces formed by the eutectic elements of the system — MUZR, LUXR, and NZXR. Let us turn the face BDC about the edge BC in

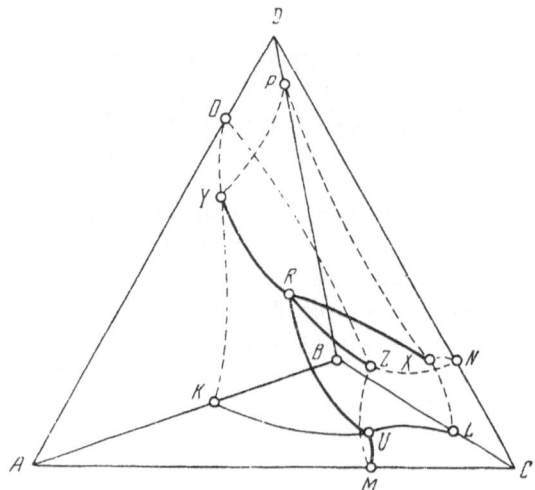

Fig. 27. Schematic representation of a quaternary system ABCD of the eutectic type, in the form of a model, by means of a tetrahedron.

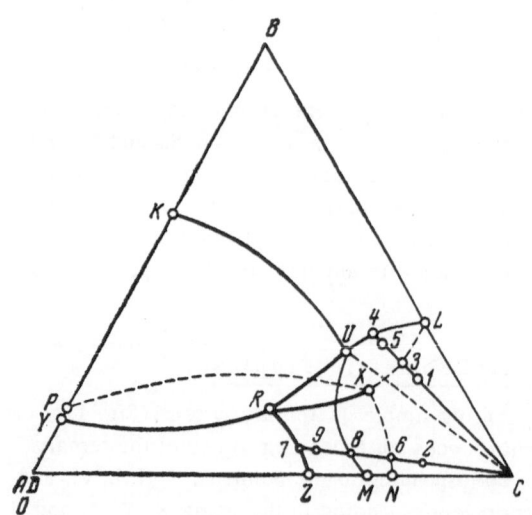

Fig. 28. Schematic representation of the quaternary system ABCD of the eutectic type, by means of the optimum projection of the tetrahedron on the plane of a diagram. Points 1, 3, 5, 4, and 2, 6, 8, 9, 7 — crystallization paths.

such a way that it coincides with the adjoining face ABC. The vertex D then coincides with the vertex A and the edges DB and DC merge with the edges AB and AC. The points O, P, N, and X then occupy the positions shown in Fig. 28.

Thus the face BDC is projected on to face ABC in its true dimensions. Here, however, the edge AD degenerates into a point, so that the face ADB (which merges with the edge AB) and the face ADC (which merges with the edge AC) become straight lines. It is evident that the ternary eutectics Y and Z will occupy definite positions on the edges AB and AC; the curved eutectic lines YO, YP, and YK, and also ZO, ZM, and ZN, are converted into sections of straight lines on the corresponding edges; finally, the entire internal volume of the tetrahedron also coincides with the face ABC. The quaternary eutectic R will therefore occupy a position inside this face (for example as shown in Fig. 28).

The point R is represented on face ABC by the concentration of components B and C and the sum (A + D). The connecting lines between the point R and the ternary eutectics U, X, Y, and Z, which in this case are projections of the corresponding curves from the internal volume of the tetrahedron on to the face ABC, are represented in exactly the same way.

As a result, of the three planar faces bounding the range of crystallization of the phase C, only two are represented on Fig. 28 — NXLC and MULC — (both in their true scales): the third planar face MNZC degenerates into a straight line. All three curved faces — MURZ, LURX, and NXRZ — are here represented in the form of planes bounded by straight and curved lines. The range of crystallization of the phase C as a whole occupies an angle at the vertex C and extends to the limits bounded by curves LXN and LURZ, depending on the relatively concentration of the components in the compositions of the system which are studied. Figure 28 also gives the range of crystallization of the phase B in the system ABCD; this obviously adjoins the vertex B and extends to the limits bounded by the curves LUK and LXRY. In order to represent the ranges of crystallization of the other two solid phases of the system — A and D — it is necessary to construct an analogous diagram by turning the face ABD about the edge AD in such a way that it coincides with the adjoining face ACD.

We constructed the phase diagram of the simple quaternary system on the basis of data on the ternary constituent systems and the quaternary eutectic. The formation of solid solutions or chemical compounds in our quaternary system would not alter the actual course of the construction. If nothing were known of the quaternary system as a whole, this would affect only the degree of accuracy of the boundaries obtained. Even if in Fig. 28 the extreme boundaries of extension of phase C in the system ABCD are unknown, we can still state confidently that its range of crystallization adjoins the vertex C and is bounded in any case by lines LXN and LUM.

The optimum projection of the triangular prism — $ABCA_1B_1C_1$ — is obtained by the superposition of two adjacent faces of the figure — ACA_1C_1 and BCB_1C_1 (Fig. 12b). Let us use this prism to represent the reciprocal system $3//2$, so that each of its vertices corresponds to one of the six simple salts of the system. The faces ACA_1C_1 and BCB_1C_1 should then correspond to the ternary reciprocal systems:

$$A + C_1 \rightleftarrows A_1 + C \text{ and } B + C_1 \rightleftarrows B_1 + C.$$

In order to obtain a construction analogous to that for the tetrahedron, it is sufficient to turn the face of the prism AA_1CC_1 about the edge CC_1 until it coincides with the adjacent face BCB_1C_1.

The two coincident faces are then also represented in their true scales and the diagram obtained indicates the boundaries of the ranges of crystallization of each of the phases formed by the salts C and C_1.

In order to represent the ranges of crystallization of the phases formed by the salts B and B_1, and also A and A_1, it is necessary to construct two other diagrams of the same optimum type, obtained in the first case by turning the face ABA_1B_1 about the edge BB_1 until it coincides with the face CBC_1B_1, and in the second case by turning the face CAC_1A_1 about the edge AA_1 until it coincides with the face BAB_1A_1.

Thus it is also possible to construct a tentative phase diagram for a quaternary reciprocal system on the basis of data on its constituent ternary (reciprocal) systems.

The Optimum Projections of Multidimensional Geometric Figures

In order to represent multicomponent systems of the first class, use is made of regular simplexes — the pentatope, hexatope, heptatope, etc. The optimum projections of the pentatope, hexatope, or any n-dimensional simplex are projections on the triangular faces of the corresponding figures.

Let us represent the five-component system ABCDE by means of a pentatope. We then have in Fig. 7 the superposition of three adjacent faces of the pentatope, corresponding to the ternary systems ADE, BDE, and CDE. This diagram can be constructed by turning two of the above faces, for example BDE and CDE, about the common edge DE until they coincide completely with the third adjacent face ADE. It is evident that none of these three faces undergo any contraction, so that all three systems are represented in their true scales. The remaining faces of the pentatope naturally undergo contraction and degenerate into points (the face ABC) or straight lines (the faces ABD, BCD, etc.).

If the phase diagrams of the ternary systems ADE, BDE, and CDE are known, their superposition in the order indicated in Fig. 7 will give an approximate picture of the boundaries of the ranges of crystallization of the phases formed by the components E and D.

In order to determine the ranges of crystallization of other phases — A, B, or C — it is of course necessary to superimpose, in corresponding fashion, other three ternary systems (with the binary system common to all of them) corresponding to other faces of the pentatope: ABC, DBC, and EBC, or DAB, EAB, and CAB.

In the six-component system ABCDEK, the tentative phase diagram can be obtained in analogous fashion as follows: for phases E and K by the superposition of the four ternary systems AEK, BKE, CKE, and DKE; and for the phases A and B or C and D by the superposition of systems ABC, ABD, ABK, and ABE, or ACD, BCD, ECD, and KCD, respectively (Fig. 13j).

Finally, for the n-component system ABCD...EF the optimum projection (Fig. 14) of the corresponding $(n-1)$-dimensional simplex makes it possible to construct diagrams representing the approximate boundaries of the ranges of crystallization of any two phases of the system, formed by any two of its components, on the basis of data on the ternary constituent systems containing these components. For example, the tentative boundaries of the ranges of crystallization of the phases formed by components A and B are obtained by the superposition and coincidence of the phase diagrams of the following $(n-2)$ ternary systems: ABC, ABD, ..., ABE, and ABF. The total number of ternary systems making up the n-component system under consideration is equal to $[(n-2)(n-1)n]/6$. Thus in each specific case use is made of only some of the ternary systems formed by the original components. Multicomponent systems of the second and third classes are represented by multidimensional geometric figures whose optimum projections are projections on their square faces. In order to represent five-component systems of the second class, use is made of the projection shown in Fig. 8b. This is obtained by the superposition, along the adjacent edge, of three square faces of the tetrahedral hexahedroid: AA_1DD_1, BB_1DD_1, and CC_1DD_1. Thus if we wished to represent the reciprocal system $4//2$ of eight salts, the above three faces would correspond to the ternary reciprocal systems: $A + D_1 \rightleftharpoons A_1 + D$; $B + D_1 \rightleftharpoons B_1 + D$; $C + D_1 \rightleftharpoons C_1 + D$. We would thus obtain the approximate boundaries of all the phases of the system formed by the salts D and D_1 (Fig. 8b). In analogous fashion, the construction of one of the tentative phase diagrams of an n-component system of the second class A, B, C, ... L, $Q//MN$ would require the superposition, by the method indicated in Fig. 16, of $(n-2)$ ternary reciprocal systems, including any two simple salts with a common cation. In this way we would obtain the approximate boundaries of the ranges of crystallization of these two salts, for example AM and AN. It should be noted that in this case the number of ternary reciprocal systems is $[(n-2)(n-1)]/2$. The simplest representatives of systems of the third class are five-component systems. They are represented by means of a prismatic hexahedroid, whose optimum projection is given in Fig. 9b. This projection is obtained by the superposition of four square faces of the original figure, each of which corresponds to one of the reciprocal ternary systems forming part of the quinary system. If the system $3//3$ includes nine simple salts: A, B, C, A_1, B_1, C_1, A_2, B_2, and C_2, the following reciprocal systems with the common salt C_2 are superimposed: $A + C_2 \rightleftharpoons A_2 + C$; $B + C_2 \rightleftharpoons B_2 + C$; $A_1 + C_2 \rightleftharpoons A_2 + C_1$; $B_1 + C_2 \rightleftharpoons B_2 + C_1$. As a result, Fig. 9b gives the approximate boundaries of the range of crystallization of the phase formed by the salt C_2 or, in the more general case, of the phases containing the salt C_2. (In Fig. 9b it is denoted AM.)

The formation of each of the nine projections of the optimum type evidently requires the combination and superposition of different quaternary systems with one common salt from the nine ternary reciprocal systems making up the quinary system.

The construction of the diagrams for n-component systems of the third class is based on the same principle. Figure 18 represents schematically the diagram of the system ABC...LQ//MNP in the range of crystallization of the phase AM formed by one of its simple salts. In this construction $2(n-3)$ ternary reciprocal systems were used; the n-component system as a whole has $[(n-3)(n-2)3]/2$ ternary reciprocal systems.

Finally, systems of any class and with any number of components $(C//A)$ are represented by means of multidimensional figures whose optimum projection is given in Fig. 21. As in the previous case, the diagram gives a clear representation of the range of crystallization of only one of the phases of the system, for example the phase containing the salt AM. If it is assumed that we have an n-component system ABC...LQ//MNP...RS, the construction of the approximate boundaries of the range of crystallization of the phase AM requires the superposition of the phase diagrams of $(n-m)(m-1)$ reciprocal ternary systems (with one common salt AM) forming part of the given system out of a total number $[(n-m)(n-m+1)(m-1)m]/4$, where $n-m+1$ is the number of cations and m the number of anions.

Thus for multicomponent systems of any type it is possible to predict the properties on the basis of data on the binary and ternary constituent systems if the method of optimum projections is used for their representation.

Tentative fusion and solubility diagrams for some systems of six or more components, obtained by this method, are given below.

FUSION DIAGRAM OF THE SIX-COMPONENT SYSTEM
Fe−Ni−Cr−Mn−Cu−Co [27]

The Fe−Ni−Cr−Mn−Cu−Co system has not been studied experimentally. As a six-component system of the first class, it includes 6 one-component, 15 binary, 20 ternary, 15 quaternary, and 6 quinary systems.

There are data in the literature only for the melting points of each of the melts Fe, Ni, Cr, Mn, Cu, and Co, and for the phase diagrams of the 15 binary and 8 of the ternary systems formed by these components.

For the construction of one of the fusion diagrams in the range rich in any two metals, it is necessary to have data on four ternary systems. Thus the total number of ternary systems on the basis of which it is possible to construct all three diagrams of the optimum type, necessary for the representation of the system as a whole, is equal to 12. These 12 ternary systems, however, should be specially selected from the 20 forming the senary system.

The question of which 12 of the 20 possible systems should be chosen in each specific case remains to some extent arbitrary and depends on the pairs into which the six components of the system being studied will be broken down.

The number of combinations of two from six is fifteen, but since the system is represented completely by means of three diagrams, the total number of possible combinations is decreased by a factor of 1/3 and is equal to 5. Figure 29 shows these five possible combinations (with 12 ternary systems in each), any of which may be chosen for the representation of the six-component system Fe−Ni−Cr−Mn−Cu−Co.

It is evident that the choice of a given combination is determined by which of the ternary systems has been most fully and effectively studied. In the Fe−Ni−Cr−Mn−Cu−Co system, the following 8 binary systems have been studied: Fe−Ni−Cr; Fe−Ni−Mn; Fe−Ni−Cu; Fe−Ni−Co; Fe−Cu−Mn; Ni−Cu−Mn; Fe−Cr−Co; Cu−Ni−Cr.

A complete fusion diagram for the system as a whole cannot be constructed on this basis. Separate regions of the system can obviously be represented, however. Moreover, the ternary systems make it possible to represent the corresponding regions of the fusion diagrams of the quaternary and quinary systems forming part of the senary system.

The method of constructing the fusion diagrams of quinary metallic systems was previously described in detail for the case of the system Fe−Ni−Cr−Mn−Cu [6]. We shall therefore give here only general information on those quinary and quaternary systems which can be represented on the basis of the fusion diagrams of the above eight ternary systems.

The Fe − Ni − Cr − Mn − Cu System

For the five quaternary systems making up this quinary system, it is possible to construct tentative fusion diagrams for all compositions only in two cases: the Fe−Ni−Cu−Mn and Fe−Ni−Cr−Cu systems. In other cases such diagrams can be constructed only for individual alloys, for example those rich in iron and nickel (in the Fe−Ni−Cr−Mn system) or in nickel and copper (in the Ni−Cu−Cr−Mn system).

Finally, in the Fe−Cu−Cr−Mn system, diagrams cannot be constructed for any compositions. For the quinary system as a whole, it is possible to construct such diagrams for alloys rich in iron, nickel, and copper.

47

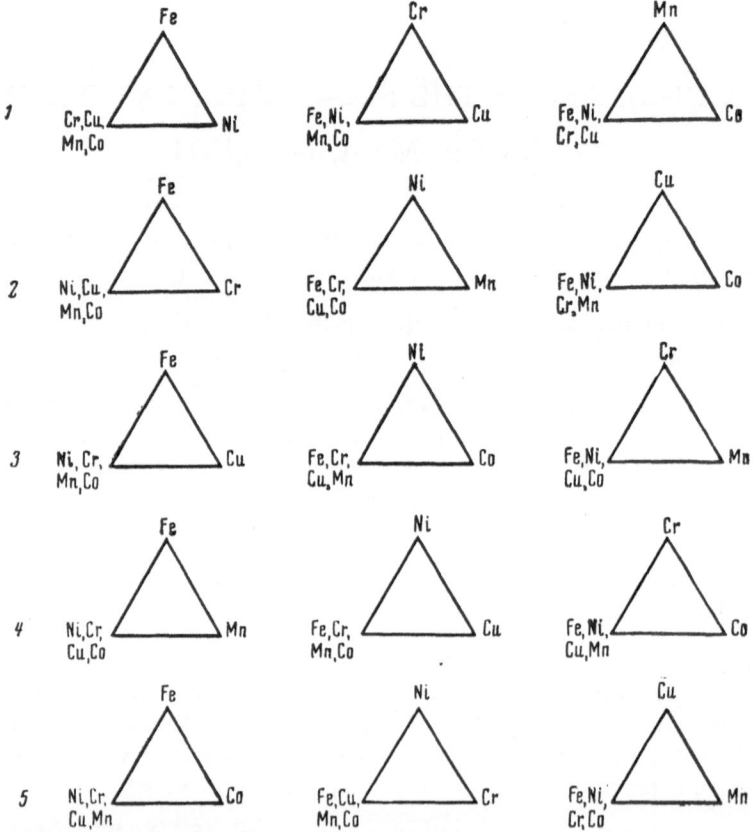

Fig. 29. Optimal projection of the hexatope representing the system
Fe−Ni−Cr−Mn−Cu−Co. Five possible combinations (1-5) of the
three fusibility diagrams are shown.

The Fe − Ni − Cu − Mn − Co System

There are slightly fewer data for this system and hence less opportunity for predicting the properties of
quaternary systems which have not been studied experimentally. In particular, for this system it is possible to
construct approximate diagrams for all compositions of the quaternary system Fe−Ni−Cu−Mn and for alloys
rich in iron and nickel in the systems Fe−Ni−Cu−Co and Fe−Ni−Mn−Co. In the other two systems Fe−Cu−
−Mn−Co and Ni−Cu−Mn−Co such constructions are in general impossible. In the quinary system as a whole
it is possible to construct diagrams for alloys rich in iron and nickel.

The Fe − Ni − Cu − Cr − Co System

Tentative fusion diagrams can be constructed for all compositions of the systems Fe−Ni−Cu−Cr and
Fe−Ni−Cr−Co, but not for any of the compositions of the systems Fe−Cu−Cr−Co and Ni−Cu−Cr−Co.

In the quaternary system Fe−Ni−Cu−Co and the quinary system Fe−Ni−Cu−Cr−Co such constructions
are possible for alloys rich in iron and nickel.

The Fe − Ni − Mn − Cr − Co System

In the quinary system Fe−Ni−Mn−Cr−Co and in the quaternary systems Fe−Ni−Mn−Cr and Fe−Ni−
−Mn−Co, tentative diagrams can be constructed for alloys rich in nickel and iron.

Constructions are possible for alloys rich in iron and chromium in the Fe−Mn−Cr−Co system and for
all compositions in the Fe−Ni−Cr−Co system: for the Ni−Mn−Cr−Co system such constructions are impossible.

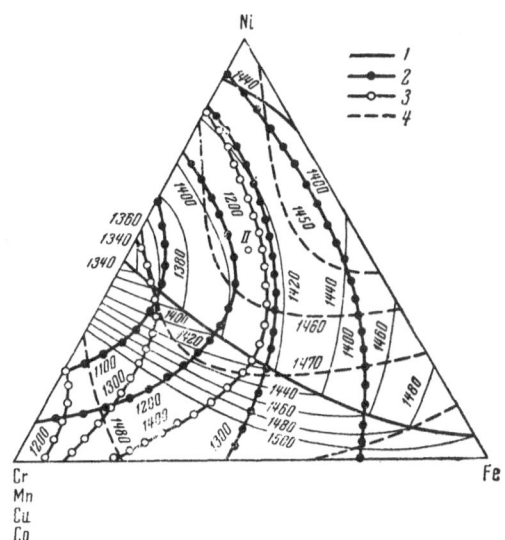

Fig. 30. Fusion diagram of the system Fe–Ni–
–Cr–Mn–Cu–Co in the range rich in iron and
nickel: 1) Ni–Fe–Cr; 2) Ni–Fe–Mn; 3)
Ni–Fe–Cu; 4) Ni–Fe–Co.

The Fe – Cu – Mn – Cr – Co System

No fusion diagrams can be constructed for the quinary system Fe–Cu–Mn–Cr–Co. This is also true of the systems Fe–Cu–Mn–Cr, Fe–Cu–Mn–Co, Fe–Cu–Cr–Co, and Cu–Mn–Cr–Co. Diagrams can be constructed only for alloys rich in iron and chromium in the Fe–Mn–Cr–Co system.

The Ni – Cu – Mn – Cr – Co System

Diagrams cannot be constructed for any compositions in the quinary Ni–Cu–Mn–Cr–Co system or the quaternary systems Ni–Cu–Mn–Co, Ni–Cu–Cr–Co, Ni–Mn–Cr–Co, and Cu–Mn–Cr–Co. Diagrams can be constructed only for alloys rich in nickel and copper in the Ni–Cu–Mn–Cr system.

Thus for two of the six quinary systems, the available data on the ternary systems do not make possible even a tentative prediction of the melting points of their compositions. For the other four, prediction is possible, but is limited for the most part to alloys based on iron and nickel or, in individual cases, copper.

At the same time allowance must be made for the fact that of the eight ternary systems which have been studied, we could use only seven for the above constructions; the eighth system Cu–Ni–Cr could not be used, since it is not included in the group of combinations of components to which the other seven belong.

When we proceed to an examination of the senary system Fe–Ni–Cr–Mn–Cu–Co, we find that the ternary systems which have been studied make it possible to construct for the senary system only one fusion diagram of the optimum type — that for alloys rich in nickel and iron (Fig. 30). This is obtained by super-position of the fusion diagrams of four ternary systems with the common binary system Ni–Fe: Ni–Fe–Cr (1), Ni–Fe–Mn (2), Ni–Fe–Cu (3), and Ni–Fe–Co (4). The composition diagram of each of the ternary systems gives isotherms corresponding to compositions with the same melting points (the isotherms for different systems are indicated by different symbols). In the representation of any six-component alloy on this diagram it is necessary to plot from below upwards the percentage concentration of nickel, and from left to right the percentage concentration of iron; all the other components (Co, Cr, Mn, and Cu) are thus represented jointly. It is evident that the regions close to the Ni and Fe vertices contain the ranges of crystallization of phases including each of these metals (or both together). The boundaries of extension of these phases are different in each of the above ternary systems. If it is assumed that in the quaternary and quinary systems and in the senary system itself the curvature of the boundary ranges is comparatively small, it is possible on the projection to record the extreme boundaries of extension of the above phases in the system as a whole. Similarly if it is assumed that in the corresponding quaternary, quinary, and senary systems the melting points of different compositions differ comparatively little from the additive values compared with the lower constituent ternary systems, the isotherms shown in Fig. 30 indicate the extreme melting points possible in the given system for its different compositions.

Let us assume, for example, that we wish to calculate the approximate melting point of an alloy rich in iron and nickel, with the following compositions: 25% Fe + 50% Ni + 3% Cr + 3% Mn + 9% Co + 10% Cu (Point II).

For convenience in the calculations, we construct two additional diagrams, on which the four super-imposed systems are combined in pairs: the Ni–Fe–Cr and Ni–Fe–Co systems in Fig. 31a, and the Ni–Fe–Cu and Ni–Fe–Mn systems in Fig. 31b. The point II in Fig. 30 is transferred to Fig. 31a and b, in such a way that the absolute concentration of iron and nickel remains unchanged.

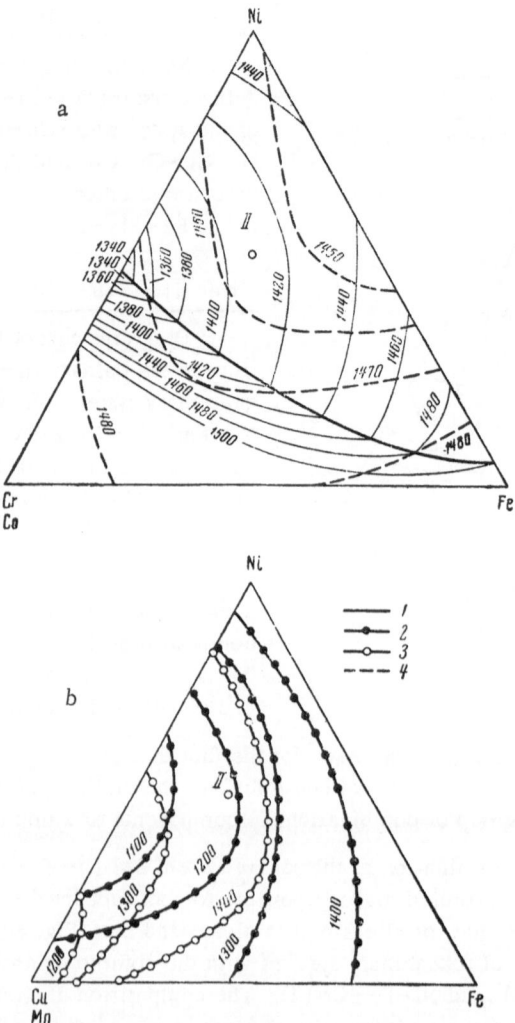

Fig. 31. Supplementary diagrams for calculating the melting points of six-component nickel–iron alloys containing
chromium, copper, manganese, and cobalt: a) 1) Ni–Fe–Cr;
4) Ni–Fe–Co; b) 2) Ni–Fe–Mn; 3) Ni–Fe–Cu.

We first determine the melting point of the composition corresponding to point II on the basis of Figs.
31a and b, respectively.

It follows from Fig. 31a that in the system Fe–Ni–Cr, i.e., at the given Fe–Ni concentration in the
absence of other components apart from Cr, the alloy corresponding to point II would melt at 1408°C; similarly,
in the Fe–Ni–Co system its melting point would be 1457°C. Since in our alloy the ratio Co:Cr = 9:3, it is
to be expected that in the quaternary Fe–Ni–Co–Cr system at the given component ratio (i.e., at absolute
concentrations of 25% Fe and 50% Ni in the absence of Cu and Mn, with the given relative concentrations of
cobalt and chromium) the melting point of the alloy would be equal to

$$1408° + \frac{(1457° - 1408°) \cdot 9}{12} = 1445°.$$

It follows from Fig. 31b that the alloy corresponding to the point II would melt at a temperature of
1240°C in the Fe–Ni–Mn system and at 1388°C in the Fe–Ni–Cu system.

Since in our alloy the ratio Cu : Mn = 10 : 3, we find by calculation that the alloy corresponding to point II in the quaternary system Fe−Ni−Mn−Cu (i.e., in the absence of cobalt and chromium) would melt at a temperature of

$$1240° + \frac{(1388° − 1240°) \cdot 10}{13} = 1354°.$$

If we compare these data and take account of the facts that the alloy in which we are interested contains six components, and that the ratio (Cu + Mn) : (Cr + Co) = 13 : 12, it can readily be seen that its melting point should be equal to approximately

$$1354° + \frac{(1445° − 1354°) \cdot 12}{25} = 1397° − 1398°.$$

Experimental verification* showed that an alloy with the composition indicated has a melting point of 1392°C. Thus in this case the calculated value is extremely close to the experimental value.

---

*The verification was carried out in the A. A. Baikov Metallurgy Institute by Candidate of Technical Sciences L. I. Pryakhina.

PREDICTION OF THE MELTING POINTS OF CERTAIN COMPOSITIONS IN THE Ni–Ti–Cr–Mo–W–Nb SYSTEM [28]

The six-component system formed by molybdenum, chromium, titanium, tungsten, niobium, and nickel is extremely complex. It contains the 15 binary, 20 ternary, 15 quaternary, and 6 quinary lower systems indicated below.

Binary Systems

1) Mo−Cr; 2) Mo−Ti; 3) Mo−W; 4) Mo−Nb; 5) Mo−Ni; 6) Cr−Ti; 7) Cr−W; 8) Cr−Nb; 9) Cr−Ni; 10)Ti−W; 11) Ti−Nb; 12) Ti−Ni; 13) W−Nb; 14) W−Ni; 15) Nb−Ni.

Ternary Systems

1) Mo−Cr−Ti; 2) Mo−Cr−W; 3) Mo−Cr−Nb; 4) Mo−Cr−Ni; 5) Mo−Ti−W; 6) Mo−Ti−Nb; 7) Mo−Ti−Ni; 8) Mo−W−Nb; 9) Mo−W−Ni; 10) Mo−Nb−Ni; 11) Cr−Ti−W; 12) Cr−Ti−Nb; 13)Cr−Ti−Ni; 14) Cr−W−Nb; 15) Cr−W−Ni; 16) Cr−Nb−Ni; 17) Ti−W−Nb; 18) Ti−W−Ni; 19) Ti−Nb−Ni; 20)W−Nb−Ni.

Quaternary Systems

1) Mo−Cr−Ti−W; 2) Mo−Cr−Ti−Nb; 3) Mo−Cr−Ti−Ni; 4) Mo−Cr−W−Nb; 5) Mo−Cr−W−Ni; 6) Mo−Cr−Nb−Ni; 7) Mo−Ti−W−Nb; 8) Mo−Ti−W−Ni; 9) Mo−Ti−Nb−Ni; 10) Mo−W−Nb−Ni; 11) Cr−Ti−W−Nb; 12) Cr−Ti−W−Ni; 13) Cr−Ti−Nb−Ni; 14) Cr−W−Nb−Ni; 15) Ti−W−Nb−Ni.

Quinary Systems

1) Mo−Cr−Ti−W−Nb; 2) Mo−Cr−Ti−W−Ni; 3) Mo−Cr−Ti−Nb−Ni; 4) Mo−Cr−W−Nb−Ni; 5) Mo−Ti−W−Nb−Ni; 6) Cr−Ti−W−Nb−Ni.

Of these, only the individual metals and the binary systems which they form have been studied in more or less detail: in addition, partial studies have been made of the six ternary systems: Ni−Cr−Mo; Ti−Mo−Nb; Ti−Cr−W; Cr−W−Mo; Ti−Cr−Mo; and Ni−Cr−W.

When we examine the extent to which the individual ternary systems making up the senary systems have been studied, we find that it is most convenient to use the method of optimum projections to construct two diagrams: one for alloys based on nickel, and the other for alloys based on titanium. In both cases we use ternary systems containing molybdenum (Figs. 32a and b).

To construct these diagrams, it is necessary to have data on the original metals and the twenty binary and seven ternary systems listed below:

1) Ni−Mo; 2) Ni−Cr; 3) Ni−W; 4) Ni−Nb; 5) Ni−Ti; 6) Mo−Cr; 7) Mo−W; 8) Mo−Nb; 9) Mo−Ti; 10)Ti−Cr; 11) Ti−W; 12) Ti−Nb.

1) Ni−Mo−Cr; 2) Ni−Mo−W; 3) Ni−Mo−Nb; 4) Ni−Mo−Ti; 5) Ti−Mo−Cr; 6) Ti−Mo−W; 7) Ti−Mo−Nb.

The original metals have the following melting points (°C): Ni − 1452, Ti − 1725, Cr − 1890, Mo − 2620, W − 3410, Nb − 2410. Almost all show polymorphic changes, the most important being those of nickel (at 358°C) and titanium (at 885°C).

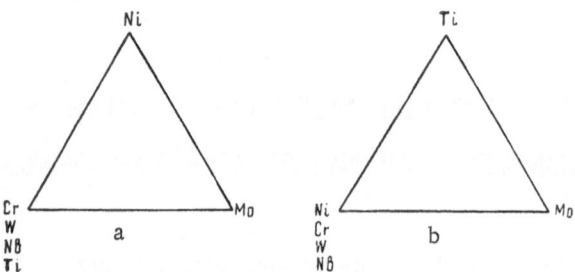

Fig. 32. Optimum projections for the representation of six-component alloys based on ternary systems containing molybdenum: a) alloys rich in nickel; b) alloys rich in titanium.

In the binary systems Ti—Nb [29], Ti—Mo [30], Mo—W [31], and Mo—Nb [31], continuous series of solid solutions are formed and the liquidus curves for these systems deviate only slightly from straight lines. A continuous series of solid solutions is also formed in the Mo—Cr system [31], but the liquidus curve has a minimum, corresponding to alloys containing 77% Cr (m.p. 1860°C).

Extensive ranges of restricted solid solutions are formed in the Ti—W [32], Ni—Cr [33], and Ti—Cr [30] systems, and also in alloys rich in nickel in the systems Ni—Mo [32], Ni—W [32], and Ni—Nb [32].

The solubility of tungsten in titanium reaches 20%, and the melting point of the alloys formed reaches 1880°C. Chromium dissolves in nickel to the extent of almost 50%, but the liquidus curve is lowered, so that the compositions with highest chromium content melt at 1325°C.

The solubility of chromium in titanium reaches more than 40% at the melting point, but decreases with decrease in temperature and amounts to only 20% at 800°C.

The dissolution of molybdenum and also of niobium in nickel is accompanied by a decrease in the melting point of the resulting alloys to 1300 and 1265°C, respectively, at a concentration of the alloying metal greater than 40%. The dissolution of tungsten in nickel is accompanied by an increase in the melting point relative to pure nickel, the value for alloys containing 40% W reaching 1505°C.

In the last three systems, cooling of alloys from the solid solutions leads to the crystallization of metallic compounds containing several nickel atoms per atom of molybdenum, niobium, or tungsten.

Finally, in the Ni—Ti system [32], the mutual solubility of the components is very low. The maximum concentration of nickel in titanium or of titanium in nickel amounts to 10%, and with decrease in temperature it decreases to 2-2.5%. Several intermetallic compounds are formed in the system (TiNi, etc.).

The remaining three binary systems forming part of the senary system are not reflected directly in Figs. 32a and b, but they also, of course, have a certain influence on the properties of the six-component alloys. It is known that in the systems Cr—W [34] and Ni—W [35], continuous series of solid solutions are formed (with a minimum, apparently, in the first case). In the Cr—Nb system [36] the mutual solubility of the components at the melting point amounts to 12.2% chromium in niobium and 16.6% niobium in chromium.

Not all of the above seven ternary systems necessary for the construction of the diagrams shown in Figs. 32a and b have been studied experimentally. Data are available in the literature only for the following.

The Ni—Cr—Mo system has been described in most detail [32, 37]. Comparison of the data obtained by different authors leads to the conclusion that solid solutions are formed in the range rich in nickel; at a nickel concentration of 60-67% these are also stable at lower temperatures. Alloys of this system based on nickel are well known for their resistance to corrosion and stability towards mineral acids (including hydrochloric).

If the chromium concentration does not exceed 15%, the alloys are fairly convenient to work with, otherwise they become brittle.

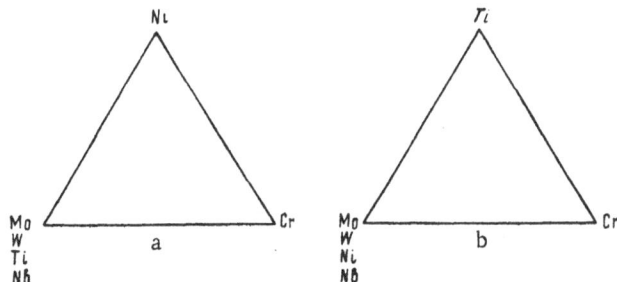

Fig. 33. Optimum projections for the representation of six-component alloys based on ternary systems containing chromium: a) alloys rich in nickel; b) alloys rich in titanium.

The Ni—W—Mo and Ni—Nb—Mo systems have not been studied experimentally, but the properties of the binary constituent systems formed by these four metals suggest that they are extremely similar and that solid solutions, stable at lower temperatures, extend over an appreciable concentration range, if the nickel concentration is not less than 60%.

The Ni—Mo—Ti system likewise has not been studied experimentally, and since the solubility of titanium in nickel is extremely low, the existence of solid solutions in the ternary system is possible only in the region of the diagram directly adjoining the nickel angle.

The ternary systems based on titanium and molybdenum have been studied in slightly more detail, particularly for compositions rich in titanium. The most detailed studies have been made of the titanium angle of the diagram in the Ti—Mo—Cr system [30] up to 35% Cr and 35% Mo. It was found that solid solutions based on α-Ti, and the compound TiCr$_2$, crystallize at higher chromium concentrations; with increase in the molybdenum concentration at the expense of chromium, however (for example, 20-30% Mo and correspondingly 15-5% Cr), only solid solutions of titanium are formed, the β-phase becoming stable at lower temperatures.

A continuous series of solid solutions was detected in the Ti—Mo—Nb system [38]. Increase in the concentration of niobium and molybdenum in the alloys is accompanied by a decrease in the temperature of the conversion of α-Ti into β-Ti; the solid solutions based on β-Ti become more stable.

The ternary system Ti—Mo—W has not been studied experimentally. All the available data on the properties of the binary constituent systems suggest, however, that it is similar to the previous system and that continuous series of solid solutions are formed. There are also some data in the literature on other ternary systems forming part of the senary system under consideration. Thus it has been found that continuous series of solid solutions are formed at all component ratios in the Cr—W—Mo system [39]. In the Ni—Cr—W system, solid solutions are formed at all ratios of chromium and molybdenum so long as the nickel concentration is not less than 60% [34]. Finally, the Ti—Cr—W system at 75% Ti contains a ternary solid solution based on titanium, irrespective of the ratio of chromium and tungsten [40].

The last three systems are not reflected directly in Figs. 32a and b. Since it is desirable to use to the maximum possible extent the available literature data and accumulated experience in the field of alloys containing the metals present in the senary system Ni—Ti—Cr—Mo—W—Nb, it is necessary to find a means of utilizing all the ternary systems which have been studied experimentally. This is particularly desirable since two of them contain chromium together with nickel or titanium. They can therefore be used to construct additional diagrams (Figs. 33a and b) which may be used to verify the data obtained from the diagrams in Figs. 32a and b. For the construction of the diagram in Figs. 33a and b it is necessary to have data not only on the binary systems but also on seven ternary systems, some of which were not used in the construction of the diagrams in Figs. 32a and b. Of these, the only ones which have been studied are: Ni—Cr—Mo; Ti—Cr—Mo; Ni—Cr—W; Ti—Cr—W. The others (Ni—Cr—Nb; Ni—Cr—Ti; Ti—Cr—Nb) have not been studied, and their fusion diagrams can be estimated only approximately, on the basis of the diagrams of the corresponding binary systems and also of some ternary systems containing related elements.

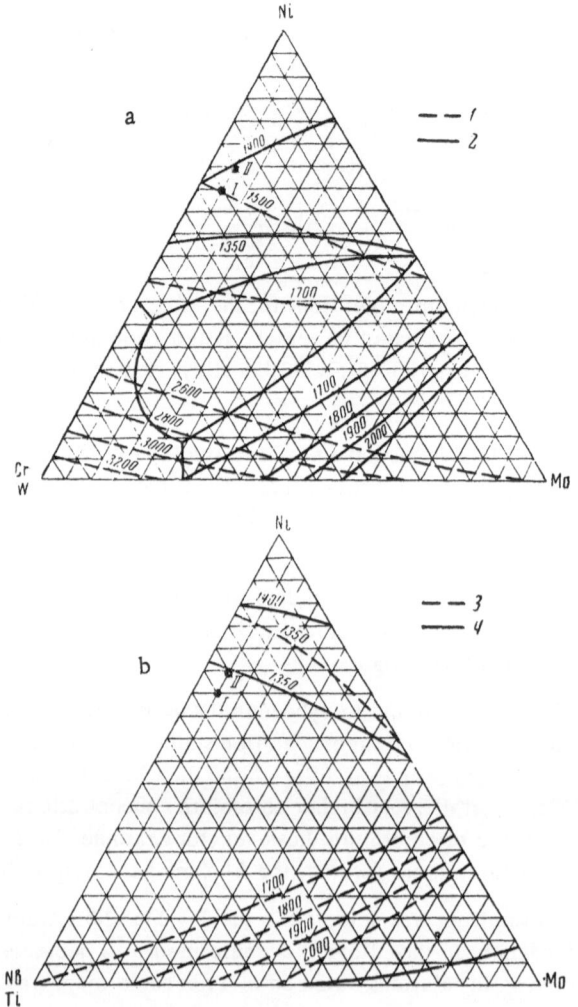

Fig. 34. Fusion diagrams for nickel alloys of the Ni—Cr—
—Ti—Mo—W—Nb system, based on ternary systems contain-
ing molybdenum: a) 1) Ni—Mo—W; 2) Ni—Mo—Cr; b) 3)
Ni—Mo—Ti; 4) Ni—Mo—Nb.

For greater convenience in the utilization of the fusion diagrams of the six-component system we shall
represent each of the diagrams mentioned by means of two parallel figures.

Thus the diagram shown in Fig. 32a is represented in Figs. 34a and b; that in Fig. 32b is represented in
Figs. 35a and b; that in Fig. 33a is represented in Figs. 36a and b; and that in Fig. 33b is represented in Figs.
37a and b. These diagrams can be used to predict the properties of six-component alloys rich in nickel and
titanium, if we make the following assumptions: 1) in systems of 4, 5, and 6 components, no new phases are
produced apart from those which are present in the constituent binary and ternary systems; and 2) as additional
components are introduced, the mutual solubility of the metals increases, and the melting point and other
properties of the multicomponent alloys change linearly and are determined by the corresponding properties
of the binary and ternary systems.

Thus Figs. 34a and b, and 36a and b are suitable for determining the properties of six-component alloys
rich in nickel; and Figs. 35a and b, and 37a and b are suitable for alloys rich in titanium. Since they were
constructed on the basis of the fusion diagrams of different ternary systems, they can be used for cross checking. Thus all the melting points of the four alloys given below were calculated on the basis of each pair of

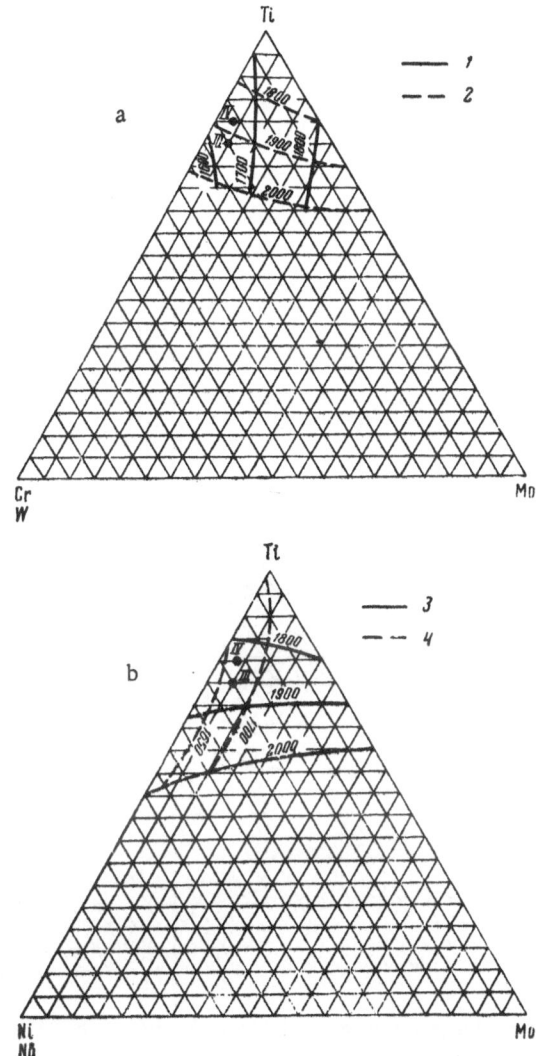

Fig. 35. Fusion diagrams for titanium alloys of the Ni−Cr−
−Ti−Mo−W−Nb system, based on ternary systems containing
molybdenum: a) 1) Ti−Mo−Cr; 2) Ti−Mo−W; b) 3) Ti−Mo−
−Nb; 4) Ti−Mo−Ni.

diagrams separately; the arithmetic mean of the data obtained was then taken. Some data on the phase
diagrams of the ternary systems used were available only for alloys rich in nickel.

The diagrams given in Figs. 34, 35, 36, and 37 were used to determine approximately the melting points
of four compositions in the senary system: 65% Ni + 15% Cr + 5% Mo + 5% Ti + 7% W + 3% Nb (alloy I); 70%
Ni + 10% Cr + 5% Mo + 5% Ti + 7% W + 3% Nb (alloy II); 75% Ti + 5% Mo + 5% W + 10% Cr + 2% Ni + 3% Nb
(alloy III); and 80% Ti + 3% Mo + 7% Cr + 5% W + 3% Nb + 2% Ni (alloy IV).

Alloy I

The values found for the melting point of this alloy from Figs. 34a and b are 1418 and 1244°C, respectively; comparison of these data with the concentrations of the components gives the value 1371°C.

Calculation of the melting point of the same alloy from Figs. 36a and b gives the values 1444 and
1225°C, respectively, or, finally, 1356°C.

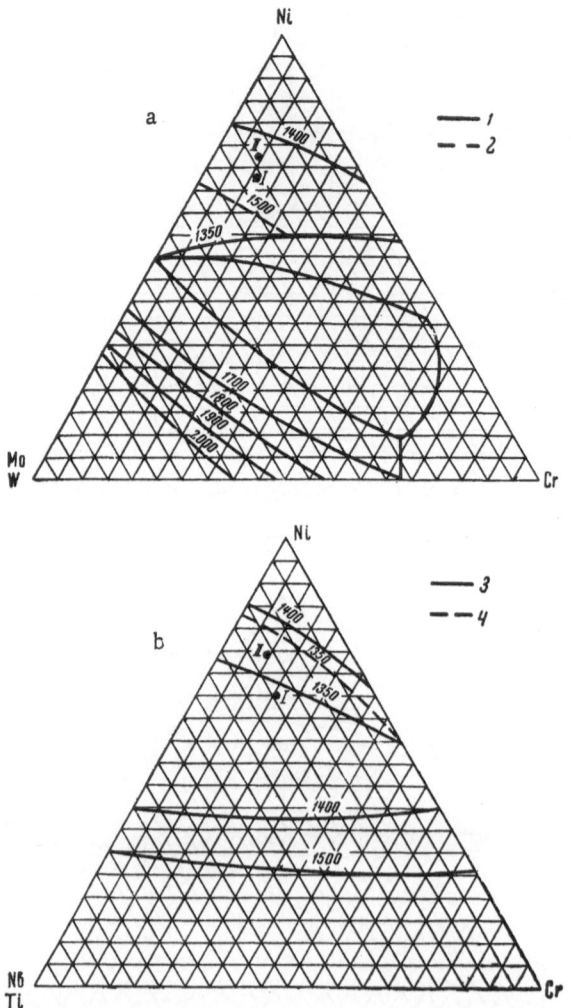

Fig. 36. Fusion diagrams of nickel alloys of the Ni—Cr—
—Ti—Mo—W—Nb system, based on ternary systems contain-
ing chromium: a) 1) Ni—Cr—Mo; 2) Ni—Cr—W; b) 3)
Ni—Cr—Nb; 4) Ni—Cr—Ti.

It may therefore be assumed that an alloy consisting of 65% Ni, 15% Cr, 5% Mo, 5% Ti, 7% W, and 3% Nb melts at 1363°C.

Experimental verification gave the value 1356°C.*

Alloy II

The values for the melting point of this alloy, calculated from Figs. 34a and b, are 1437 and 1272°C, respectively. When the ratio of the concentrations of the components is taken into account, we obtain the value 1384°C. The values obtained for the same alloy from Figs. 36a and b are 1445 and 1256°C, or, finally, 1369°C.

Thus the expected melting point of the alloy consisting of 70% Ni, 10% Cr, 5% Mo, 5% Ti, 7% W, and 3% Nb is 1376°C. Experimental verification gave the value 1350°C.*

---

*The experiments on the determination of the melting points of alloys I and II were carried out in the A. A. Baikov Metallurgy Institute by Candidate of Technical Sciences L. I. Pryakhina.

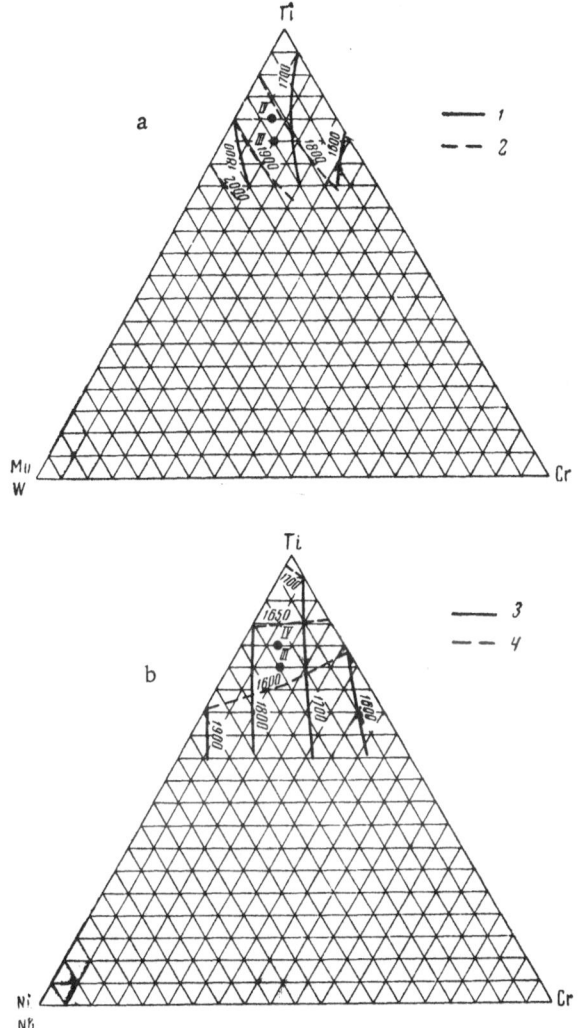

Fig. 37. Fusion diagrams of titanium alloys of the Ni—Cr—
—Ti—Mo—W—Nb system, based on ternary systems contain-
ing chromium: a) 1) Ti—Cr—Mo; 2) Ti—Cr—W; b) 3)
Ti—Cr—Nb; 4) Ti—Cr—Ni.

Alloy III

The values for the melting point of this alloy, calculated from Figs. 35a and b, are 1735 and 1786°C or,
when the ratio of the concentrations of all the components is taken into account, 1748°C. The values obtained
by calculation based on Figs. 37a and b are 1797 and 1699°C, or, finally, 1764°C. The melting point of the
alloy consisting of 75% Ti, 5% Mo, 5% W, 10% Cr, 2% Ni, and 3% Nb can thus be taken as approximately 1756°C.

Alloy IV

The values for the melting point of this alloy, calculated from Figs. 35a and b, are 1744 and 1765°C,
or, finally, 1749°C; the values found for the same alloy from Figs. 37a and b are 1800 and 1708°C, respectively,
or, finally, 1764°C. It may therefore be assumed that an alloy containing 80% Ti, 3% Mo, 7% Cr, 5% W,
3% Nb, and 2% Ni melts at approximately 1757°C.

The melting point of compositions of this system, rich in titanium, can apparently be raised by increasing the relative concentration of tungsten and niobium at the expense of nickel and chromium. The fusibility
of its molybdenum, niobium, and more particularly tungsten alloys is even higher.

Approximate data for alloys based on molybdenum can be obtained from the same Figs. 34a and b and 35a and b. Since the calculations are based on the fusion diagrams of different ternary systems forming part of the senary system (in the first case they include molybdenum and nickel with a third component, and in the second case they include molybdenum and titanium with a third component), they can be used for cross checking and to obtain more accurate data.

The following are the approximate melting points of some alloys rich in molybdenum, calculated by the method used for the corresponding alloys of nickel and titanium:

60% Mo + 30% Cr + 3% Ti + 2% W + 3% Nb + 2% Ni — m.p. 2059°C; 50% Mo + 30% Cr + 4% Ti + 10% W + 4% Nb + 2% Ni — m.p. 2032°C; 50% Mo + 30% Cr + 8% Ti + 4% W + 6% Nb + 2% Ni — m.p. 2000°C; 45% Mo + 30% Cr + 7% Ti + 9% W + 4% Nb + 5% Ni — m.p. 1973°C; 40% Mo + 38% Cr + 6% Ti + 7% W + 4% Nb + 5% Ni — m.p 1844°C; 30% Mo + 60% Cr + 3% Ti + 2% W + 3% Nb + 2% Ni — m.p. 1624°C.

PROVISIONAL 25 °C SOLUBILITY ISOTHERM FOR THE TEN-COMPONENT SYSTEM Ba, In, Be, Na, Cd, Co, Cu, Li//Cl, Br + H_2O

In the representation of multicomponent reciprocal systems containing a solvent, it is most convenient to confine attention to the salt composition, representing the water (or other solvent) in the form of isolines on the composition diagram. The dimensionality of the geometric figure by means of which the system should be represented is then reduced by one, since one of the components is omitted. Thus in the case under consideration it is possible to omit the water and to select for the representation of the nine-component salt composition the eight-dimensional analog of the tetrahedral hexahedroid. This system as a whole, and its lower constituent nine-, eight-, seven-, six-, five-, and even four-component systems have not been studied experimentally. The literature contains only a few data on the solubility of individual salts and some ternary systems. At the same time the Ba, In, Be, Na, Cd, Co, Cu, Li//Cl, Br + H_2O system contains 16 binary aqueous salt systems (corresponding to the 16 simple salts), 64 ternary, 140 quaternary, 196 quinary, and 182 senary systems, and also 112, 44, and 10 seven-, eight-, and nine-component systems — a total of 764 systems of different classes (simple and reciprocal).

It was pointed out above that for the construction of tentative diagrams it is possible to confine attention to experimental data on ternary and in some cases even binary constituent systems, and that if we represent not the system as a whole but only the ranges of crystallization of its individual solid phases, the number of these ternary and other constituent systems need not be too large. For the construction of the tentative isotherm of our nine-component system in the range of crystallization of one of its simple salts it is necessary to have the corresponding data on only 22 ternary systems (since we are restricting ourselves to the representation of the salt composition, they correspond to the edges of the figure). These 22 systems, however, must be specially selected in order to satisfy the requirements arising from the conditions of formation of the optimum projection of the eight-dimensional figure used for the representation of the salt composition of our system. Thus, for example, the isotherm in the range of crystallization of barium chloride can be constructed on the basis of the solubility isotherms of the following systems:

1) $BaCl_2 - BaBr_2 - H_2O$; 2) $BaCl_2 - LiCl - H_2O$; 3) $BaCl_2 - CdCl_2 - H_2O$; 4) $BaCl_2 - BeCl_2 - H_2O$; 5) $BaCl_2 - NaCl - H_2O$; 6) $BaCl_2 - CoCl_2 - H_2O$; 7) $BaCl_2 - CuCl_2 - H_2O$; 8) $BaBr_2 - LiBr - H_2O$; 9) $BaBr_2 - CdBr_2 - H_2O$; 10) $CdCl_2 - CdBr_2 - H_2O$; 11) $BaBr_2 - NaBr - H_2O$; 12) $BaBr_2 - CuBr_2 - H_2O$; 13) $BaBr_2 - CoBr_2 - H_2O$; 14) $BaCl_2 - InCl_3 - H_2O$; 15) $BaBr_2 - BeBr_2 - H_2O$; 16) $BaBr_2 - InBr_3 - H_2O$; 17) $NaCl - NaBr - H_2O$; 18) $BeCl_2 - BeBr_2 - H_2O$; 19) $InCl_3 - InBr_2 - H_2O$; 20) $CuCl_2 - CuBr_2 - H_2O$; 21) $CoCl_2 - CoBr_2 - H_2O$; 22) $LiCl - LiBr - H_2O$.

The isotherms for systems 1-10 have been studied and described in the literature [41]. We have calculated the isotherms for systems 11-14 on the basis of the law of mass action [42] from the solubility of the simple salts, since it was possible to assume that no chemical compounds or solid solutions are formed. For two of the systems (15 and 16) even these calculated isotherms could not be obtained because of the absence of data on the solubility of $BeBr_2$ and $InBr_3$ in water. In view of the high solubility of the bromides of all the other metals in our complex system, and also by analogy with the ternary systems containing beryllium or indium chloride, barium chloride, and water, we assumed that the eutonic points in the last two systems correspond approximately to the same gram-equivalent ratios of the components as in the system $BaBr_2 - LiBr - H_2O$. Finally, for systems 17-22 there were no data, apart from indirect references. The solubility isotherms of the systems $BaCl_2 - BaBr_2 - H_2O$ and $CdCl_2 - CdBr_2 - H_2O$ suggest that solid solutions are formed in all cases in aqueous solutions of the chloride and bromide of the same metal.

TABLE 8. Salt Composition of Components and Water Content at the Eutonic Points of
Some Ternary Systems at 25 °C

System	Comp. of solution, equiv. %		H_2O, g per g eq of salt	Solid phase
	Component I	Component II		
$BaCl_2 — BaBr_2 — H_2O$ [43]	100	0	281.2	$BaCl_2 \cdot 2H_2O$
	9.1	90.9	144.7	Solid solution I + solid solution II
	0	100	149	$BaBr_2 \cdot 2H_2O$
$BaCl_2 — LiCl — H_2O$ [44]	1.8	98.2	79	$BaCl_2 \cdot 2H_2O + LiCl \cdot H_2O$
	0	100	50.2	$LiCl \cdot H_2O$
$BaCl_2 — CdCl_2 — H_2O$ [43]	63.8	36.2	148.4	$BaCl_2 \cdot 2H_2O + BaCl_2 \cdot CdCl_2 \cdot 4H_2O$
	38.4	61.6	113	$BaCl_2 \cdot CdCl_2 \cdot 4H_2O + BaCl_2 \cdot 2CdCl_2 \cdot 5H_2O$
	5.2	94.8	72	$BaCl_2 \cdot 2CdCl_2 \cdot 5H_2O + CdCl_2 \cdot 2,5H_2O$
	0	100	75	$CdCl_2 \cdot 2,5H_2O$
$BaCl_2 — BeCl_2 — H_2O$ [45]	0.4	99.6	52	$BaCl_2 \cdot 2H_2O + BeCl_2 \cdot 4H_2O$
	0	100	51.6	$BeCl_2 \cdot 4H_2O$
$BaCl_2 — NaCl — H_2O$ [46]	7.2	92.8	162	$BaCl_2 \cdot 2H_2O + NaCl$
	0	100	168	$NaCl$
$BaCl_2 — CoCl_2 — H_2O$ [47]	0.8	99.2	121	$BaCl_2 \cdot 2H_2O + CoCl_2 \cdot 6H_2O$
	0	100	117	$CoCl_2 \cdot 6H_2O$
$BaCl_2 — CuCl_2 — H_2O$ [48]	3.9	96.1	84	$BaCl_2 \cdot 2H_2O + CuCl_2 \cdot 2H_2O$
	0	100	85	$CuCl_2 \cdot 2H_2O$
$BaBr_2 — LiBr — H_2O$ [49]	100	0	140	$BaBr_2 \cdot 2H_2O$
		100	52	$BaBr_2 \cdot 2H_2O$
	0	100	47	$LiBr \cdot 2H_2O$
$BaBr_2 — CdBr_2 — H_2O$ [43]	54.5	45.5	63	$BaBr_2 \cdot 2H_2O + BaBr_2 \cdot CdBr_2 \cdot 4H_2O$
	44.4	55.6	60.4	$BaBr_2 \cdot CdBr_2 \cdot 4H_2O + CdBr_2$
	37.2	62.8	73	$CdBr_2 + CdBr_2 \cdot 4H_2O$
	0	100	122	$CdBr_2 \cdot 4H_2O$
$CdCl_2 — CdBr_2 — H_2O$ [43]	64.6	35.4	73.5	Solid solution of $CdCl_2 \cdot 2,5H_2O$ and $CdBr_2 \cdot 4H_2O$
$BaBr_2 — NaBr — H_2O$	24.5	75.5	62.8	$BaBr_2 \cdot 2H_2O + NaBr \cdot 2H_2O$
	0	100	111	$NaBr \cdot 2H_2O$
$BaBr_2 — CuBr_2 — H_2O$	29.3	70.7	69	$BaBr_2 \cdot 2H_2O + CuBr_2$
	0	100	113	$CuBr_2$
$BaBr_2 — CoBr_2 — H_2O$	27	73	60	$BaBr_2 \cdot 2H_2O + CoBr_2 \cdot 6H_2O$
	0	100	92.6	$CoBr_2 \cdot 6H_2O$
$BaCl_2 — InCl_3 — H_2O$	0.7	99.3	30	$BaCl_2 \cdot 2H_2O + InCl_3 \cdot 4H_2O$
	0	100	33	$InCl_3 \cdot 4H_2O$

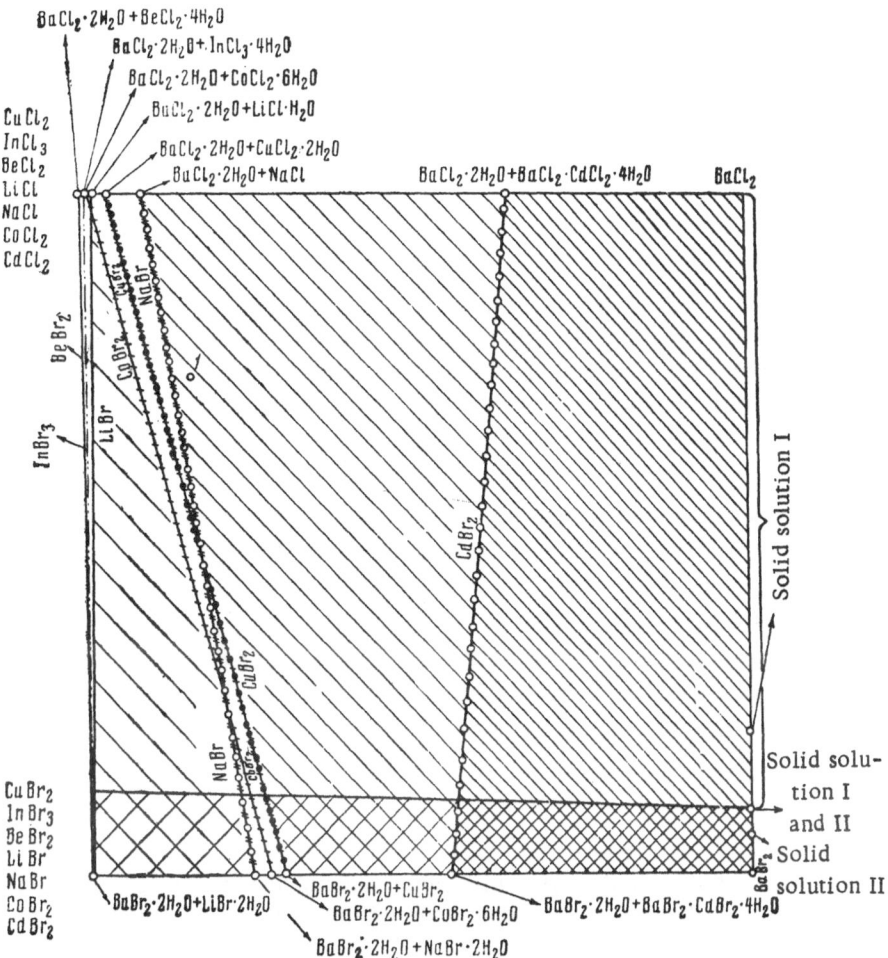

Fig. 38. Range of crystallization of $BaCl_2 \cdot 2H_2O$ in the ten-component system
Ba, Be, In, Li, Co, Cu, Na, Cd//Cl, Br + H_2O at 25°C.

The absence of more definite data for these six systems cannot however, present a serious obstacle to the solution of the problem under consideration, since the corresponding components are represented jointly on the optimum projection.

Thus the provisional 25°C isotherm for the ten-component system in the range of crystallization of $BaCl_2 \cdot 2H_2O$ (Fig. 38) is constructed using the data given in Table 8.

Figure 38 gives the provisional boundaries of the range of crystallization of barium chlorides and bromides in the system Ba, In, Be, Na, Cd, Co, Cu, Li//Cl, Br + H_2O at 25°C.

Unlike the case of similar diagrams for ternary reciprocal systems, in this case only the most probable minimum and maximum limits of these ranges are indicated. The curves indicate the salt which forms with $BaCl_2$ the most stable reciprocal salt pair. Water is not shown, but it can be calculated from Table 8, which gives the quantity of water for saturated solutions, at 25°C, of the original simple salts at the eutonic points of the ternary systems.

The isotherm given in Fig. 38 enables us to reach a number of conclusions. It is quite obvious that in this system, crystalline barium chloride cannot be obtained in pure form at any component ratio, since even traces of bromide ion are sufficient for the formation of solid solutions of barium chloride and bromide.

Two forms of solid solution are possible: those based on $BaCl_2 \cdot 2H_2O$ and $BaBr_2 \cdot 2H_2O$. The first type of solid solution (which we shall arbitrarily refer to as "barium chloride") exists up to relatively high concentrations of bromide in the solution — up to 90% in the sum of the anions: the second type of solid solution (referred to arbitrarily as "barium bromide") exists at relatively low concentrations of chlorides — not greater than 10% in the sum of the anions.

It then follows (Fig. 38) that the range of crystallization of "barium chloride" from aqueous solutions at 25°C in the presence of the chlorides and bromides of indium, beryllium, sodium, cadmium, cobalt, copper, and lithium has minimum and maximum limits.

The minimum limits are bounded by the $CdBr_2$ line, i.e., by a high concentration of cadmium ions in the solution relative to the sum of all the other cations with the exceptions of barium. The maximum limits of the range of crystallization of "barium chloride" extend to the $BeBr_2$, LiBr, and $InBr_3$ lines, i.e., in the presence of predominating quantities of beryllium, indium, and lithium in the total cations. In the first case, for the isolation of "barium chloride," it is necessary to have in the cation composition from 56 to 65% of half Ba^{2+} ions; in the second case the presence in the cation composition of from 2 to 6% of half Ba^{2+} ions is sufficient.

Finally, in the case where the copper, cobalt, and sodium salts predominate in the solution, the range of crystallization of "barium chloride" occupies intermediate values and this salt begins to crystallize when the sum of the cations contains from 15% (in an excess of chlorides) to 30% (in an excess of bromides) of half Ba^{2+} ions.

Let us assume that we have a solution with the following composition per 100 g of water: 26.05 g $BaCl_2$ (0.25 g-eq) + 11.45 g $CdCl_2$ (0.125 g-eq) + 6.07 g $CoBr_2$ (0.055 g-eq) + 10.55 g $BeBr_2$ (0.125 g-eq) + 3.46 g $InCl_3$ (0.031 g-eq) + 5.14 g NaBr (0.1 g-eq) + 5.31 g LiCl (0.250 g-eq) + 6.72 g $CuCl_2$ (0.1 g-eq) (point 1 on the diagram).

At point 1 we have in the sum of the anions 27% Br^- and 73% Cl^-, so that the chloride content is quite sufficient for the crystallization of barium chloride. The barium ion content in the sum of the cations amounts to 24.13%, i.e., it lies far beyond the limits corresponding to the minimum quantity required for the crystallization of $BaCl_2 \cdot 2H_2O$.

If the solution contained no salts of beryllium, lithium, and indium, it may be assumed on the basis of the ratio $Cd : (Na + Co + Cu) = 0.125 : 0.255$ that the boundary of the field of crystallization of barium chloride would lie close to the NaBr line.

Since the solution also contains these very readily soluble salts, and $(Li + Be + In) : (Cd + Na + Co + Cu) = 0.406 : 0.370$, the boundary of the field of crystallization of $BaCl_2 \cdot 2H_2O$ should be displaced even further to the left on approximately the same horizontal as point 1. It may therefore be assumed that the solution taken is in equilibrium with a solid phase consisting of a solid solution of barium chloride and bromide, based on $BaCl_2 \cdot 2H_2O$.

THE RIVER SYSTEM AND PROBLEMS OF THE
CLASSIFICATION OF NATURAL WATERS

In the salt composition of river and ground water, the chlorides, sulfates, and bicarbonates of sodium, calcium, and magnesium usually predominate. The six-component system Na^+, Ca^{2+}, $Mg^{2+}//Cl^-$, SO_4^{2-}, $HCO_3^- + H_2O$ has therefore been called the river system. Because of the particular importance of fresh waters for the national economy, a considerable amount of work has been devoted to their study. Various classification principles have been proposed, to provide some direction in a vast amount of experimental material. Among these, an important place is taken by geometric methods, based partly on certain principles of physicochemical analysis.

Professor S. A. Durov was the first to propose that the salt composition of fresh waters be represented in the form of a "double triangular" diagram" [50]. It was shown that this diagram consists of a set of three planar projections of a prismatic hexahedroid, proposed earlier by V. P. Radishchev [51]. A. G. Bergman later adopted the same method for the representation of the river system, supplementing it by certain data in accordance with the classification characteristics of fresh waters.

In both cases the basis of the classification was provided by the salt composition, which, other conditions being equal, determines which of the salts present in the river system crystallizes first when the corresponding solution becomes saturated.

It is evident that the solution of this problem, irrespective of the method of representation, requires a precise knowledge of the solubility polytherm of the river system.

Unfortunately, however, no such experimental studies have as yet been made. The literature contains only incomplete data on the solubility of the individual salts, some ternary systems, and a very few quaternary reciprocal systems forming part of the senary system as a whole.

As pointed out above, the method of optimum projections makes it possible not only to represent rationally systems which have already been the subject of a complete experimental study, but also to construct albeit tentative solubility diagrams even in cases where only a few compositions have been studied experimentally, for example where several ternary reciprocal systems are known. Thus by using the data available in the literature it is possible to construct tentative solubility isotherms (25°C) for the ranges of crystallization of the follow-

TABLE 9. Solubility of Salts of the River System in Water at 25°C

Salt	Concentration in solution		Solid phase	Salt	Concentration in solution		Solid phase
	wt.%	mol. per 1000 g H_2O			wt.%	mol. per 1000 g H_2O	
2NaCl	26.3	3.05	NaCl	$Ca(HCO_3)_2$	0.145	0.009	$CaCO_3$
Na_2SO_4	21.75	1.97	$Na_2SO_4 \cdot 10H_2O$	$MgCl_2$	35.54	5.79	$MgCl_2 \cdot 6H_2O$
$2NaHCO_3$	9.39	0.61	$NaHCO_3$	$MgSO_4$	26.68	3.03	$MgSO_4 \cdot 7H_2O$
$CaCl_2$	45.06	7.58	$CaCl_2 \cdot 6H_2O$	$Mg(HCO_3)_2$	0.233	0.016	$MgCO_3 \cdot 3H_2O$
$CaSO_4$	0.209	0.015	$CaSO_4 \cdot 2H_2O$				

TABLE 10. Composition of the Eutonic Points of Ternary Systems Forming the River System, at 25°C

System	Comp. of solution, mol. per 1000 g H_2O		Salt comp. of solution, equiv.%		Solid phase
	Component I	Component II	Component I	Component II	
$Ca(HCO_3)_2$ — $Mg(HCO_3)_2$ — H_2O	0.00250	0.01126	18.17	81.83	$CaCO_3 \cdot MgCO_3 + MgCO_3$
	0.00810	0.00118	87.28	12.72	$CaCO_3 \cdot MgCO_3 + CaCO_3$
$Ca(HCO_3)_2$ — $CaSO_4$ — H_2O	0.00752	0.01295	36.74	63.26	$CaCO_3 + CaSO_4 \cdot 2H_2O$
$Mg(HCO_3)_2$ — $MgSO_4$ — H_2O	0.1840	3.0610	5.67	94.33	$MgSO_4 \cdot 7H_2O + MgCO_3 \cdot 3H_2O$
$CaSO_4$ — $MgSO_4$ — H_2O	0.00389	3.1149	0.12	99.88	$CaSO_4 \cdot 2H_2O + MgSO_4 \cdot 7H_2O$
$MgCl_2$ — $MgSO_4$ — H_2O	5.56	0.52	91.5	8.5	$MgCl_2 \cdot 6H_2O + MgSO_4 \cdot 6H_2O$
	4.45	0.51	89.7	10.3	$MgSO_4 \cdot 6H_2O + MgSO_4 \cdot 7H_2O$
$CaCl_2$ — $MgCl_2$ — H_2O	6.73	1.91	77.9	22.1	$CaCl_2 \cdot 6H_2O + CaCl_2 \cdot 2MgCl_2 \cdot 12H_2O$
	5.18	2.80	64.9	35.1	$MgCl_2 \cdot 6H_2O + CaCl_2 \cdot 2MgCl_2 \cdot 12H_2O$
$MgCl_2$ — $NaCl$ — H_2O	5.73	0.13	97.8	2.2	$MgCl_2 \cdot 6H_2O + NaCl$
Na_2SO_4 — $CaSO_4$ — H_2O	3.61	0.006	99.834	0.166	$Na_2SO_4 + CaSO_4 \cdot Na_2SO_4$
	2.46	0.02	99.193	0.807	$CaSO_4 \cdot 2H_2O + CaSO_4 \cdot Na_2SO_4$
Na_2SO_4 — $MgSO_4$ — H_2O	1.33	2.71	32.9	67.1	$MgSO_4 \cdot 7H_2O + MgSO_4 \cdot Na_2SO_4 \cdot 4H_2O$
	1.98	2.01	49.7	50.3	$Na_2SO_4 \cdot 10H_2O + MgSO_4 \cdot Na_2SO_4 \cdot 4H_2O$
$NaHCO_3$ — Na_2SO_4 — H_2O	0.33	1.92	14.7	85.3	$NaHCO_3 + Na_2SO_4 \cdot 10H_2O$
$NaCl$ — Na_2SO_4 — H_2O	1.75	1.44	54.8	45.2	$Na_2SO_4 \cdot 10H_2O + Na_2SO_4$
	2.76	0.71	79.5	20.5	$Na_2SO_4 + NaCl$
$NaHCO_3$ — $NaCl$ — H_2O	0.10	3.05	3.1	96.9	$NaHCO_3 + NaCl$
$CaCl_2$ — $NaCl$ — H_2O	7.06	0.16	97.8	2.2	$CaCl_2 \cdot 6H_2O + NaCl$
$NaHCO_3$ — $Ca(HCO_3)_2$ — H_2O	0.61	0.00013	99.98	0.02	$NaHCO_3 + CaCO_3$
$NaHCO_3$ — $Mg(HCO_3)_2$ — H_2O	0.608	0.00045	99.93	0.07	$NaHCO_3 + MgCO_3 \cdot 3H_2O$
$CaSO_4$ — $CaCl_2$ — H_2O	0.0007	7.37	0.01	99.99	$CaSO_4 \cdot 2H_2O + CaCl_2 \cdot 6H_2O$
$Ca(HCO_3)_2$ — $CaCl_2$ — H_2O*	0.00005	7.363	0.0007	99.9993	$CaCO_3 + CaCl_2 \cdot 6H_2O$
$MgCl_2$ — $Mg(HCO_3)_2$ — H_2O*	5.7	0.0002	99.996	0.004	$MgCl_2 \cdot 6H_2O + MgCO_3 \cdot 3H_2O$

*Calculated eutonic point, obtained on the assumption that no double salts or solid solutions are formed in this system.

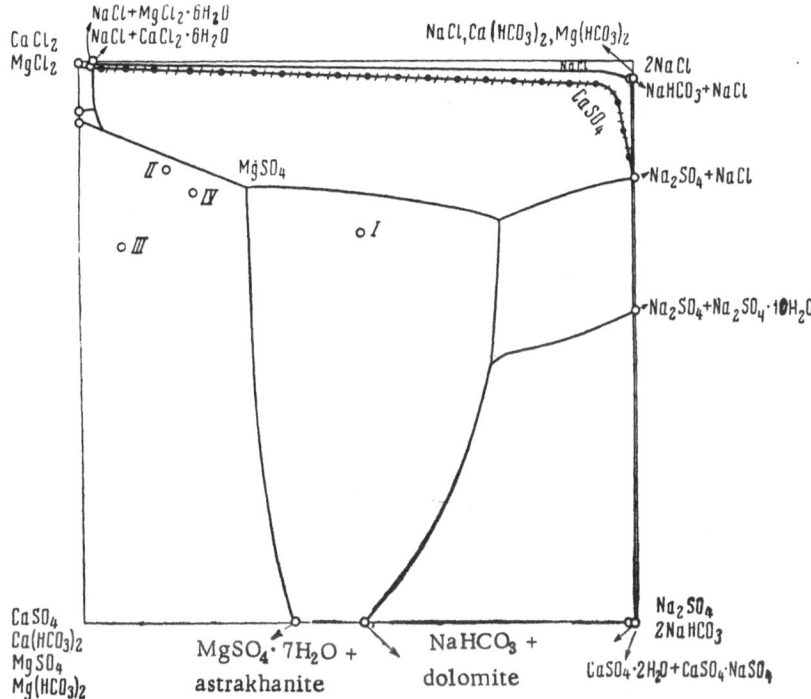

Fig. 39. Range of crystallization of NaCl in the river system at 25°C. I-IV) Ground waters.

ing simple salts present in the river system: $NaCl*$; $Na_2SO_4 \cdot 10H_2O$; $NaHCO_3$; $CaSO_4 \cdot 2H_2O$; $MgCO_3 \cdot 3H_2O$; $CaCO_3$; $MgSO_4 \cdot 7H_2O$, and also the double salts which they form — astrakhanite, glauberite, and mixed carbonates of calcium and magnesium (dolomite under natural conditions). The salt $MgCl_2 \cdot 6H_2O$ and $CaCl_2 \cdot 6H_2O$ (and also their double salt tachhydrite) occupy such negligibly small ranges of crystallization (because of the high solubility of these salts compared with the other salts of the river system) that it is extremely improbable that they will crystallize first from fresh waters. The corresponding diagrams are therefore not considered further.

The basis for the constructions given below was provided by data on the solubility at 25°C in the binary and ternary aqueous salt systems (Tables 9 and 10). They were supplemented by data in the literature on the following quaternary reciprocal systems: Na^+, $Mg^{2+}//Cl^-$, $SO_4^{2-} + H_2O$ [52]; Na^+, $Ca^{2+}//Cl^-$, $SO_4^{2-} + H_2O$ [53]; Na^+, $Ca^{2+}//SO_4^{2-}$, $HCO_3^- + H_2O$ [54]; Ca^{2+}, $Mg^{2+}//SO_4^{2-}$, $HCO_3^- + H_2O$; [55]; Na^+, $Ca^{2+}//Cl^-$, $HCO_3^- + H_2O$ [56]; Mg^{2+}, $Ca^{2+}//Cl^-$, $SO_4^{2-} + H_2O$ [41].

The range of crystallization of NaCl is given in Fig. 39. The separation of this salt requires a comparatively high relative concentration of chloride ions. Thus if bicarbonate, and also calcium salts, predominate in the solution, it is necessary that chloride ions make up not less than 90-95% in the total anions if sodium chloride is to separate. Even when magnesium sulfates predominate greatly, NaCl can crystallize only if the Cl^- content is of the order of 80%.†

Figure 40 shows the minimum and maximum boundaries of the ranges of crystallization of sodium sulfate decahydrate and anhydrous sodium sulfate, and its double salt with magnesium sulfate and four molecules of water of crystallization (astrakhanite). For the separation of these salts in the presence of an excess of chlorides from solutions of low calcium content, it is sufficient to have 25-30% magnesium ions and not more than 20% SO_4^{2-} ions. If calcium is present in appreciable quantities, however, together with bicarbonates, it becomes improbable that sodium sulfate will precipitate first from the solution, since this requires not less than 99% sodium ions in the original solution.

---

*The range of crystallization of this salt in the system Na, Mg, Ca//Cl, SO_4, $HCO_3 + H_2O$ at 25°C was derived earlier [6].

†All the calculations are here given in g-eq, and the percentages are calculated separately for the total cations and total anions (see Table 10).

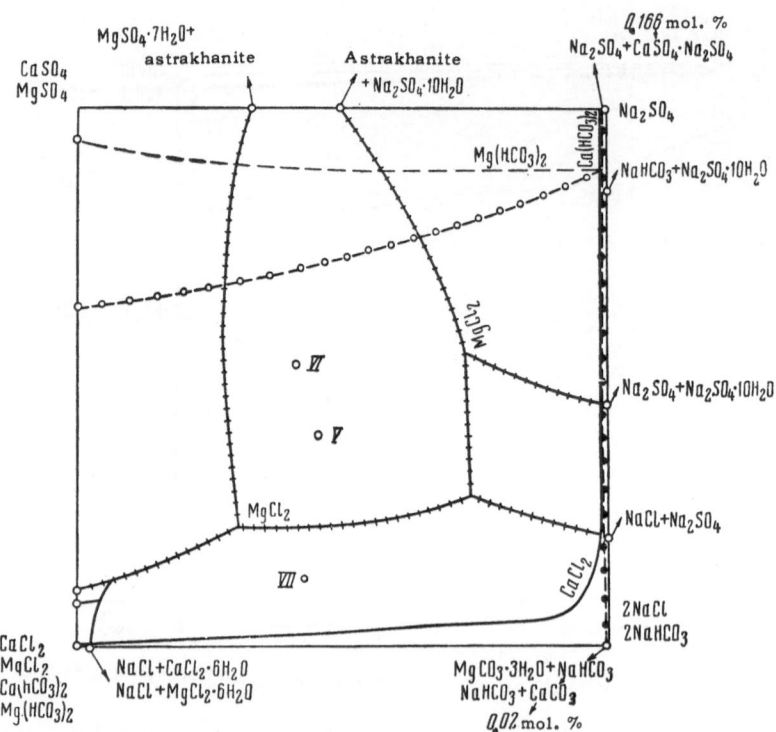

Fig. 40. Range of crystallization of $Na_2SO_4 \cdot 10H_2O$ in the river system
at 25°C. V-VII) Ground waters.

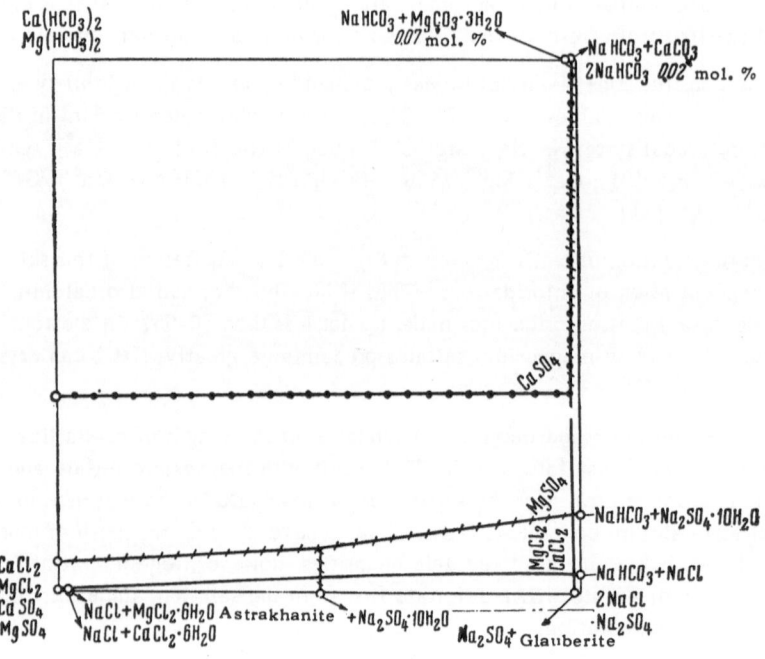

Fig. 41. Range of crystallization of $NaHCO_3$ in the river system at 25°C.

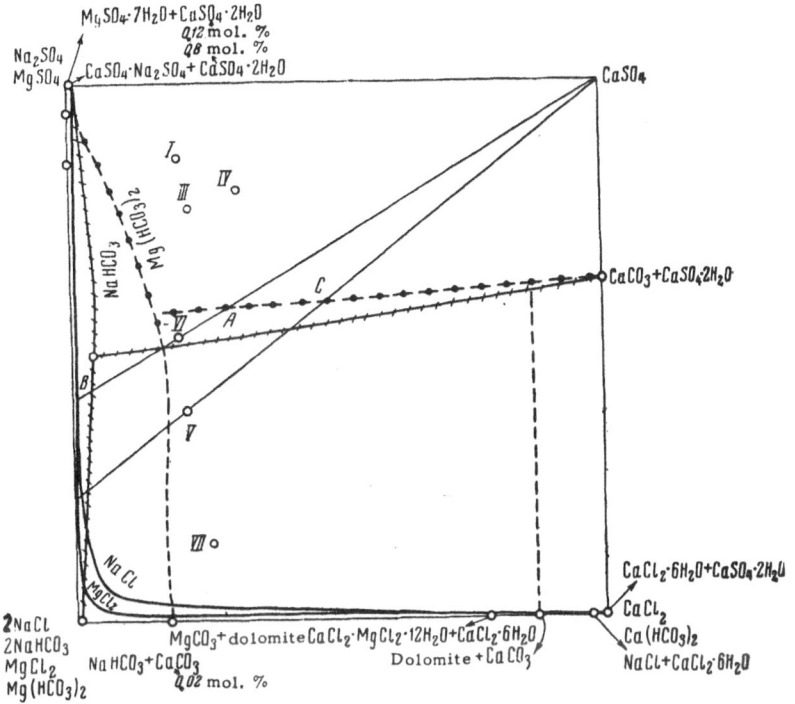

Fig. 42. Range of crystallization of $CaSO_4 \cdot 2H_2O$ in the river system at 25°C.
I, III, IV) Ground waters; V-VII) internal waters.

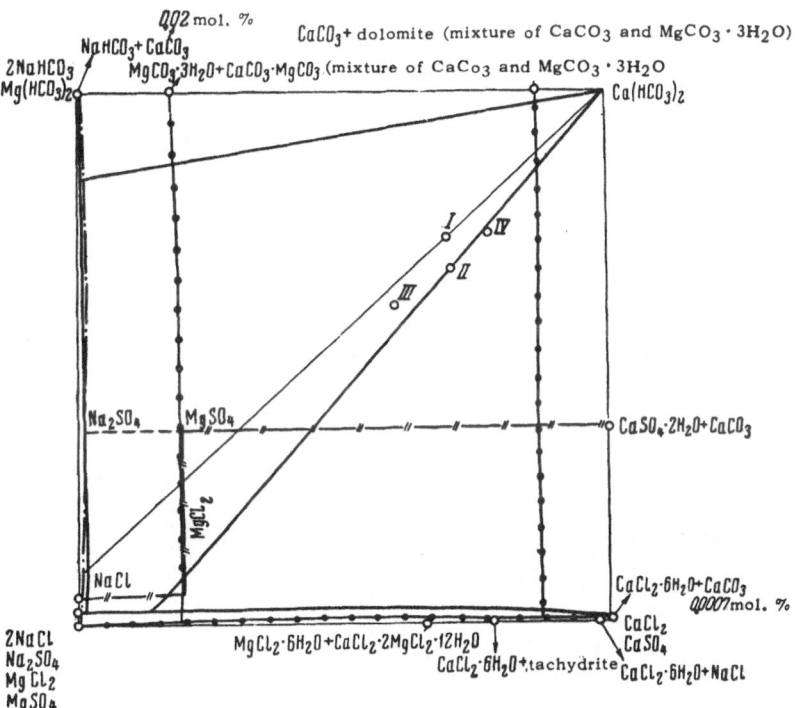

Fig. 43. Range of crystallization of $CaCO_3$ in the river system at 25°C. I-IV) River waters.

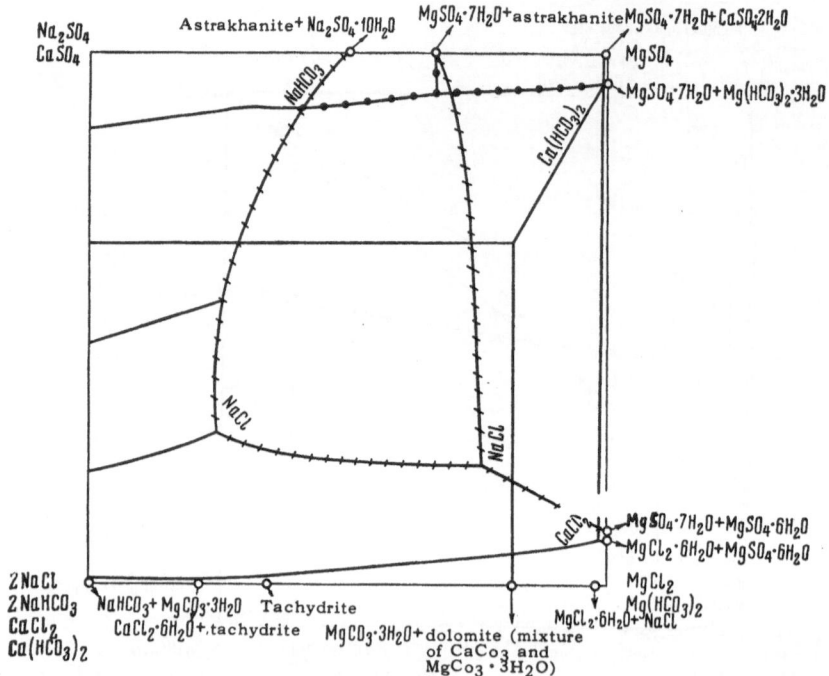

Fig. 44. Range of crystallization of $MgSO_4 \cdot 7H_2O$ and $MgSO_4 \cdot Na_2SO_4 \cdot 4H_2O$
in the river system at 25°C.

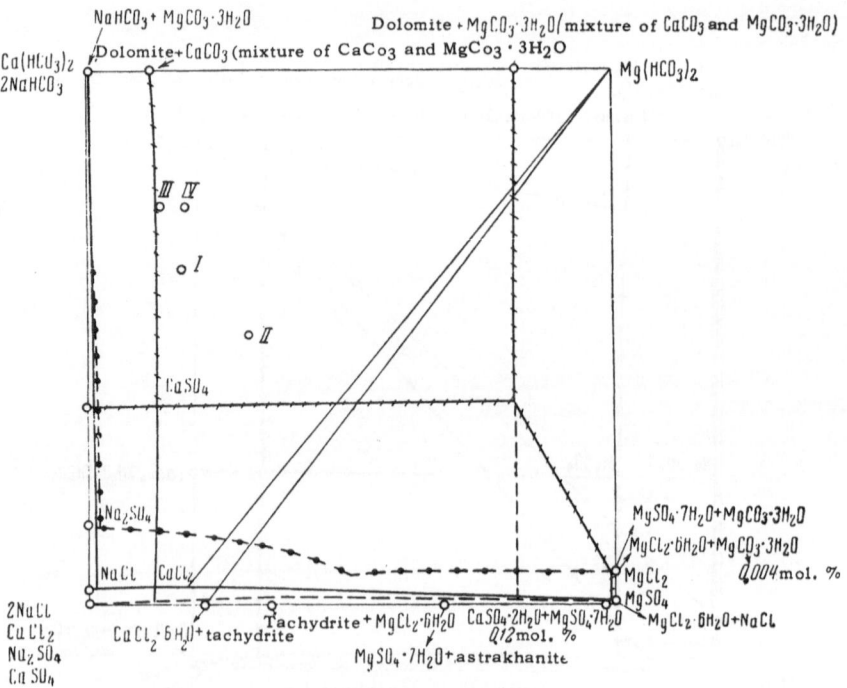

Fig. 45. Range of crystallization of $MgCO_3 \cdot 3H_2O$ in the river system at 25°C.
I-IV) River waters.

The range of crystallization of sodium bicarbonate is shown in Fig. 41. Under all conditions, it occupies a narrow band adjoining the edge of the square corresponding to the sodium salts. In other words, the crystallization of $NaHCO_3$ from fresh waters is unlikely, since it is possible only at a very high relative concentration of sodium ions (of the order of 99%), irrespective of the anion ratio.

Although in some cases only a small quantity of bicarbonate ions is required ($\sim$ 5-6%), it is nevertheless difficult to count on their presence at the moment when the calcium and magnesium contents become so low.

The boundaries of the range of crystallization of gypsum are shown in Fig. 42. The dimensions of this range are appreciable. The probability of separation of this salt is particularly high for solutions containing much chloride. A high concentration of magnesium ions reduces it to only a slight extent, and it is only in the presence of an excess of bicarbonate that the crystallization of $CaSO_4 \cdot 2H_2O$ requires not less than 60% SO_4^{2-} in the sum of the anions. In this case the presence of 3-5% calcium ions in the total cations is sufficient.

In the salt composition of the river system, the range of crystallization of gypsum is exceeded only by the range of crystallization of calcium carbonate, separately or in its mixture with magnesium carbonate.

It follows from Fig. 43 that the separation of mixed calcium and magnesium carbonates from fresh water takes place even at concentrations of only 2-3% HCO_3^- in the total anions in the salt composition of the solution, in the presence of 2-20% Ca^{2+} in the total cations (depending on the relative quantities of the other components of the river system).

The range of crystallization of epsomite ($MgSO_4 \cdot 7H_2O$) and astrakhanite ($MgSO_4 \cdot Na_2SO_4 \cdot 4H_2O$) is given in Fig. 44. If the solution contains a large quantity of calcium salts, the crystallization of magnesium sulfate is extremely unlikely — the range of crystallization of this salt is very small and lies close to the $MgSO_4$ vertex on the diagram. On the other hand, if sodium salts predominate greatly in the solution, the field of crystallization of epsomite and, more particularly, astrakhanite becomes fairly extensive *; for this it is sufficient that the Mg^{2+} ions amount to 20-40% in the sum of the cations. Under these conditions a large excess of sulfates is not required.

Finally, the range of crystallization of magnesium carbonate, separately and in its mixture with calcium carbonate (Fig. 45) is also very extensive, particularly if the original solution contains a relatively small amount of sulfates.

The diagrams given in Figs. 39-45 make possible not only qualitative but also quantitative calculations for waters of the river system.

S. A. Durov's monograph [50] gives detailed analytical data on the composition of various river, ground, and mineral waters of the USSR. Some of these were plotted on diagrams by the method of optimum projections, and the following results were obtained. It was found that in a number of cases it is possible to determine correctly which salt crystallizes first from given waters, i.e., the type to which they belong. This possible when the figurative point of the salt composition lies on one of the diagrams of Figs. 39-45 in the minimum range of crystallization of the given salt. A negative result is equally reliable: if the corresponding figurative point lies outside the limits of the maximum range of crystallization of the given salt, it obviously cannot separate first. Thus it was found that, as expected, natural waters most frequently lie in the ranges of crystallization of gypsum and calcium and magnesium carbonates.

For example, the ground and other internal waters of the Northern Caucasus lie for certain in the field of crystallization of $CaSO_4 \cdot 2H_2O$ (Fig. 42, all points except V-VII). From the waters of large rivers such as Volga, Don, Northern Dvina, and Kuban', however, mixed carbonates of calcium and magnesium undoubtedly crystallize first (Figs. 43 and 45, points I-IV). Figure 39 gives an example of a negative answer to the question asked. It is evident that points I-IV, representing certain ground waters of the Primorsk region with a high concentration of chlorides, do not lie in the range of crystallization of NaCl.

---

*Allowance must be made for the fact that, although natural deposits of astrakhanite are fairly thick, under artificial conditions, for the case of evaporation of solutions, astrakhanite is a metastable phase and crystallizes with difficulty.

There are cases, however, when the question of the nature of various fresh waters can be answered only with greater or lesser probability on the basis of these diagrams. Examples include waters with compositions corresponding to points V-VII in Figs. 40 and 42. They occupy a position intermediate between the minimum and maximum fields of crystallization of gypsum or a mixture of calcium and magnesium carbonates. Only from Fig. 40 does it follow that crystallization of astrakhanite is possible from solutions V and VI. Waters with composition VII do not belong to this type, and it may be assumed that in this case crystallization of gypsum is most probable.

The following procedure may be recommended for the determination of the nature and class of natural fresh waters whose chemical analysis is unknown.

The data obtained by chemical analysis are expressed in the form of the salt composition, calculated in gram-equivalents and as percentages of the total cations and total anions, respectively, and the composition obtained is plotted on the diagrams of Figs. 39-45. This reveals which salt (of those represented on these diagrams) is most likely to crystallize. When the answer obtained is not definite enough, the method of calculation described above must be used (pp. 49, 50, 50—60).

Since these calculations are based on the assumption that the change in the range of crystallization of the corresponding phase in the multicomponent system with change in composition is additive, they are tentative in character and the conclusions reached should be verified and made more precise by experimental methods.

PHASE DIAGRAM OF THE SEVEN-COMPONENT SYSTEM
K, Na, Li, Tl//F, Cl, NO_3, SO_4 IN THE RANGE OF
CRYSTALLIZATION OF POTASSIUM CHLORIDE

A seven-component system of the fourth class contains 16 one-component systems, 48 binary, 68 ternary (including 32 of the first class and 36 of the second), 56 quaternary (including 8 of the first class and 48 of the second), 28 quinary (including 12 of the second class and 16 of the third), and 8 senary systems of the third class.

Thus the system K, Na, Li, Tl// F, Cl, NO_3, SO_4 has 224 lower constituent systems of different classes. In order to construct its tentative phase digram in the range of crystallization of KCl, however, it is sufficient to have data on the following 9 ternary reciprocal systems:

1) K, Na//Cl, F; 2) K, Li//Cl, F; 3) K, Tl//Cl, F; 4) K, Na//Cl, NO_3; 5) K, Li//Cl, NO_3; 6) K, Tl//Cl, NO_3; 7) K, Na//Cl, SO_4; 8) K, Li//Cl, SO_4; 9) K, Tl//Cl, SO_4.

The literature contains more or less detailed data on only seven of these [57]: for two systems — K, Li//Cl, F and K, Tl//Cl, F there are no data, even on two of the boundary binary systems: TlCl—TlF and KF—TlF.

The prediction of the boundaries of the range of crystallization of potassium chloride in this system can thus be only approximate.

The aim of the constructions is rather to illustrate the actual method as applied to more complex cases when the total number of components is large and the class of the system higher than the third.

Tables 11-13 give the main characteristics, taken from literature data, of the fusion diagrams of the binary and ternary systems forming part of the seven-component system being studied. The diagram (Fig. 46) gives the boundaries of the range of crystallization of KCl in each of the above nine ternary reciprocal systems, whose phase diagrams are superimposed with a common vertex corresponding to KCl. It may thus be assumed that the range of crystallization of this salt in the seven-component system as a whole extends within a certain range from the minimum dimensions which it occupies in the system K, Na//Cl, F (approximately 13% of the area of the projection) to the maximum dimensions which it occupies in the system K, Li//Cl, NO_3 (approximately 50% of this area).

It is evident that the higher the relative concentration of Na^+ in the total cations and the relative concentration of F^- in the total anions, for corresponding compositions of the system, the smaller the range of crystallization of KCl. The dimensions of this range are also comparatively small in the case where thallium salts or, finally, thallium and sulfates, predominate in the composition in addition to fluorides.

TABLE 11. Melting Points of Simple Salts of the Na, K, Li, Tl//F, Cl, NO_3, SO_4 System

Salt	M.p., °C	Salt	M.p., °C	Salt	M.p., °C	Salt	M.p., °C
KCl	775	LiCl	606	NaCl	800	TlCl	429
KNO_3	337	$LiNO_3$	255	$NaNO_3$	308	$TlNO_3$	206
K_2SO_4	1069	Li_2SO_4	856	Na_2SO_4	884	Tl_2SO_4	632
KF	850	LiF	847	NaF	990	TlF	—

73

TABLE 12. Binary Systems Bounding the Range of Crystallization of KCl in the Seven-Component System K, Na, Li, Tl//F, Cl, NO$_3$, SO$_4$

System	Nature of singular point	Comp. of singular point, equiv.%		M.p., °C	Solid phase
		Component I	Component II		
KCl — NaCl	Minimum	50	50	658	Solid solution mKCl·nNaCl
KCl — KF	Eutectic	55	50	606	KCl + KF
NaCl — NaF	»	66.5	33.5	675	NaCl + NaF
KF — NaF	»	60	40	716	Two solid solutions mKF·nNaF
KCl — KNO$_3$	»	18	82	360	KCl + KNO$_3$·KCl
KCl — KNO$_3$	»	6	94	320	KNO$_3$ + KNO$_3$·KCl
NaCl — NaNO$_3$	Minimum	6.5	93.5	297	NaCl + NaNO$_3$
KNO$_3$ — NaNO$_3$	Minimum	50	50	223	Solid solutions mKNO$_3$·nNaNO$_3$
KCl — K$_2$SO$_4$	Eutectic	58	42	690	KCl + K$_2$SO$_4$
NaCl — Na$_2$SO$_4$	»	35	65	628	NaCl + Na$_2$SO$_4$
K$_2$SO$_4$ — Na$_2$SO$_4$	Minimum	18	82	1100	Solid solutions mK$_2$SO$_4$·nNa$_2$SO$_4$
KCl — LiCl	Eutectic	42	58	348	KCl + LiCl
KF — LiF	»	50	50	492	KF + LiF
LiCl — LiF	»	72	28	498	LiCl + LiF
KNO$_3$ — LiNO$_3$	»	57	43	125	KNO$_3$ + LiNO$_3$
LiCl — LiNO$_3$	»	11.8	88.2	252	LiCl + LiNO$_3$
K$_2$SO$_4$ — Li$_2$SO$_4$	Dystectic	57.5	42.5	726	K$_2$SO$_4$ + K$_2$SO$_4$·Li$_2$SO$_4$
K$_2$SO$_4$ — Li$_2$SO$_4$	Eutectic	50.0	50.0	734	K$_2$SO$_4$·Li$_2$SO$_4$
K$_2$SO$_4$ — Li$_2$SO$_4$	»	20.0	80.0	538	Li$_2$SO$_4$ + K$_2$SO$_4$·Li$_2$SO$_4$
LiCl — Li$_2$SO$_4$	»	47.5	52.5	480	LiCl + Li$_2$SO$_4$
KCl — TlCl	»	7.5	92.5	426	KCl + TlCl
TlCl — TlNO$_3$	»	19.5	80.5	178	TlCl + TlNO$_3$
KNO$_3$ — TlNO$_3$	»	31	69	182	Solid solutions mKNO$_3$·nTlNO$_3$
K$_2$SO$_4$ — Tl$_2$SO$_4$	»	3	97	629	K$_2$SO$_4$ + Tl$_2$SO$_4$
TlCl — Tl$_2$SO$_4$	»	37	63	358	TlCl + Tl$_2$SO$_4$
KF — TlF	—	—	—	—	—
TlCl — TlF	—	—	—	—	—

TABLE 13. Ternary Reciprocal Systems Bounding the Range of Crystallization of KCl
in the Seven-Component System K, Na, Li, Tl//F, Cl, NO₃, SO₄

System	Nature of singular point	Composition of singular point, equiv. %				M.p., °C	Solid phase
		cations		anions			
		I	II	I	II		
K, Na//Cl, F	Minimum	50.5	49.5	90.75	9.25	612	NaF + solid solution mKCl·nNaCl
K, Na//Cl, F	Eutectic	90.5	9.5	51	49	582	KCl + two solid solutions mKF·nNaF
K, Na//Cl, F	Maximum	72	28	72	28	660	KCl + NaF
K, Li//Cl, SO₄	Eutectic	37.5	62.5	83	17	349	KCl + LiCl + Li₂SO₄
K, Li//Cl, SO₄	»	35.6	64.4	23.3	76.7	443	Li₂SO₄+KCl+K₂SO₄·Li₂SO₄
K, Li//Cl, SO₄	»	75.5	24.5	26	74	575	KCl+K₂SO₄+K₂SO₄·Li₂SO₄
K, Tl//Cl, NO₃	Transition	91	9	26	74	317	KCl + TlCl + KCl·KNO₃
K, Tl//Cl, SO₄	Eutectic	10.5	89.5	97.5	2.5	414	KCl + TlCl + K₂SO₄
K, Na//Cl, NO₃	Transition	89.5	10.5	12	88	285	KCl + KNO₃ + KCl·KNO₃
K, Na//Cl, NO₃	»	76.5	23.5	10.5	89.5	244	KCl + KNO₃ + NaCl
K, Na//Cl, SO₄	Eutectic	39	61	41.5	58.5	514	KCl + NaCl + solid solution mK₂SO₄·nNa₂SO₄
K, Li//Cl, NO₃	Maximum	31.8	68.2	31.8	68.2	450	KCl + LiNO₃
K, Li//Cl, F	—	—	—	—	—	—	
K, Tl//Cl, F	—	—	—	—	—	—	

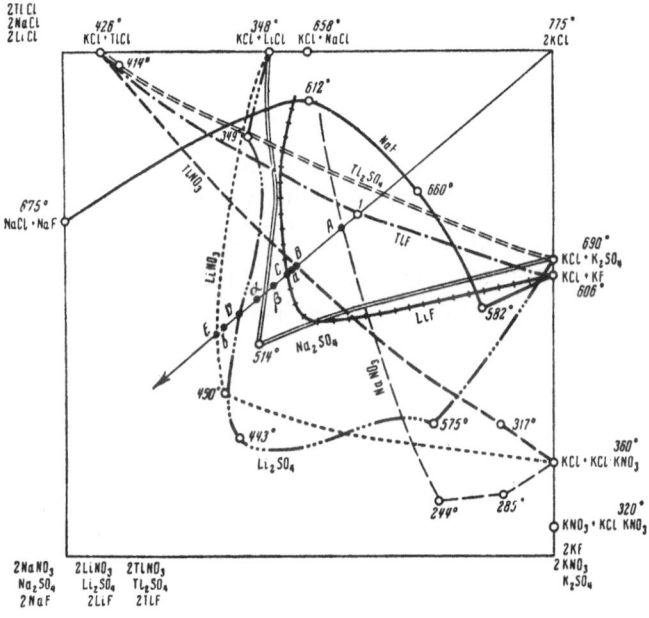

Fig. 46. Boundaries of the range of crystallization of KCl
in the system K, Na, Li, Tl//F, Cl, NO₃, SO₄.

With increase in the quantity of lithium and, simultaneously, nitrates or sulfates, the range of crystallization of KCl approaches the maximum limits. From a knowledge of the relative concentration of all components for each specific mixture of salts, we can readily use Fig. 46 to calculate whether potassium chloride can crystallize from the mixture, and if so at what temperature.

Let us assume, for example, that we have the seven-component salt mixture: 60% KCl + 7.5% NaCl + 2.5% $NaNO_3$ + 8.5% LiF + 5% $TlNO_3$ + 4% Li_2SO_4 + 12.5% $LiNO_3$ (composition expressed in equiv. %). In other words, the total cations contain 60% K^+, 5% Tl^+, 10% Na^+, and 25% Li^+ and the total anions contain 67.5% Cl^-, 8.5% F^-, 20% NO_3^-, and 4% SO_4^{2-}. In Fig. 46 this mixture is represented by the point 1. It can readily be seen that it lies inside the range of crystallization of KCl. This conclusion follows from the following considerations. In the case of the crystallization of potassium chloride from this salt mixture, the composition of the liquid melt is displaced along a ray joining the point 1 to the vertex of the square corresponding to 100% KCl, in the direction indicated by the arrow. This ray intersects the boundary eutectic lines between the fields of crystallization of KCl and the other salts in the equilibrium in the ternary reciprocal systems at the points A, B, C, D, and E. This indicates, naturally, the points of intersection of the KCl crystallization ray with the eutonic lines of the ternary systems only in those cases where the composition corresponding to the point 1 contained salts forming with KCl a reciprocal pair on a stable diagonal.

The points B and C correspond to $TlNO_3$ and LiF, which are present in the mixture in the ratio 5:8.5, so that they can be replaced by the point a.

Similarly the joint action of the salts $LiNO_3$ and Li_2SO_4, whose ratio is equal to 12.5:4, leads to the replacement of the points D and E by the point b. Since the ratio of the two salt sums ($TlNO_3$ + LiF):($LiNO_3$ + Li_2SO_4) = 13.5:16.5, the center of gravity between the points a and b will be the point α. Finally, since the salt sum ($TlNO_3$ + LiF + $LiNO_3$ + Li_2SO_4) in our mixture amounts to 30%, and the $NaNO_3$ to 2.5%, we find the boundary of the range of crystallization of KCl for the salt mixture corresponding to the point 1 — at the point β. Only the influence of NaCl, whose concentration is equal to 7.5%, has not been taken into account. The only apparent result of its influence is that a solid solution of potassium and sodium chlorides separates (appreciably enriched in potassium). Thus when a molten mixture of salts, with the above composition, is cooled, potassium chloride crystallizes (with a small quantity of sodium chloride in solid solution) until the concentration of KCl in the liquid phase decreases from 60 to 42% (i.e., the crystallization path moves from point 1 to point β). At what temperature does the mixture at point 1 melt and at what temperature does the mixture at point β solidify? The diagram given in Fig. 46 can be used to answer this question, albeit only approximately.

The melting point of the melt at the final crystallization β is evidently approximately 435°C, so that the mixture of salts (corresponding to point 1 on the diagram) will melt at about 548°C.

It is possible in analogous fashion to estimate, for any composition of the given system, whether it lies in the range of crystallization of KCl, and (if the answer is positive) to determine its approximate melting point. If the data on the phase diagrams of the ternary reciprocal systems, necessary for the construction of the optimum projections of the K, Na, Li, Tl//F, Cl, NO_3, SO_4 system in the ranges of crystallization of its other simple salts (KF, $LiNO_3$, TlCl, etc.) were available, it would be possible to predict the melting points of any seven-component salt mixture from the given components.

SOLUBILITY POLYTHERMAL FOR THE SYSTEM
K^+, Na^+ // Cl^-, SO_4^{2-}, $Cr_2O_7^{2-}$ – H_2O

The polythermal crystallization of salts from solutions is one of the basic operations in chemical technology. It usually takes place in multicomponent systems which are also reciprocal systems. Potassium bichromate is also obtained under industrial conditions by polythermal crystallization from aqueous solutions in the presence of sodium chlorides and sulfate. The five-component reciprocal system

$$K^+, Na^+ // Cl^-, SO_4^{2-}, Cr_2O_7^{2-} - H_2O$$

is formed.

Since crystallization in this case may lead to the precipitation of not only potassium bichromate but also a number of other salts, correct control of the process requires a knowledge of suitable diagrams indicating the dependence of the composition of the solid phases on the concentration of the components and temperature.

Methods for constructing phase diagrams for systems of this type are not well developed. The geometric figures by means of which they are represented have no planar projections suitable for quantitative calculations. Thus the representation of the solubility polythermal of a five-component reciprocal system is possible (if we restrict ourselves to the salt composition) by means of a prismatic heptahedroid (Fig. 47). Analysis has shown, however, that this figure has no optimum projections in coordinate space.

It therefore becomes necessary to select a suitable model, since a three-dimensional projection is generally more readily understood than a two-dimensional one, thanks to the use of the third dimension.

We are considering the K^+, Na^+ // Cl^-, SO_4^{2-}, $Cr_2O_7^{2-}$ – H_2O system from the viewpoint of the polythermal crystallization of potassium bichromate. We shall therefore restrict ourselves to the construction of a model which is the optimum three-dimensional projection for the volumes of crystallization of the bichromates, i.e., potassium bichromate and sodium bichromate.

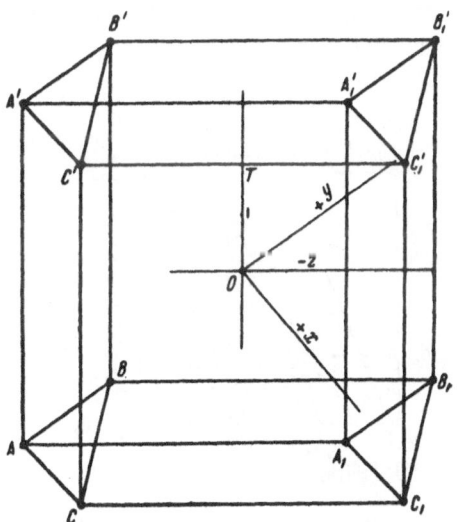

Fig. 47. Prismatic heptahedroid.

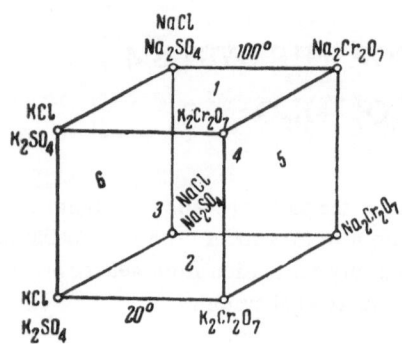

Fig. 48. Optimum projection of the heptahedroid in coordinate space. One of the solubility polythermals (20-100℃) for the system Na^+, $K^+//Cl^-$, SO_4^{2-}, $Cr_2O_7^{2-}-H_2O$.

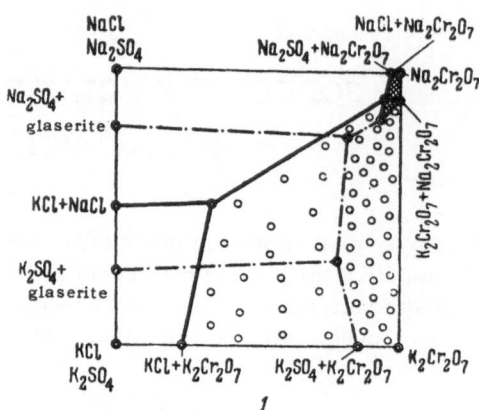

Fig. 49. Solubility isotherm (100℃) for the system Na^+, $K^+//Cl^-$, SO_4^{2-}, $Cr_2O_7^{2-}-H_2O$; salt composition in the field of crystallization of $K_2Cr_2O_7$.

We shall assume that the polythermal crystallization is carried out in the temperature range 20-100℃. We then have to construct the model shown in Fig. 48, i.e., a cube whose faces correspond to the salt compositions of the following systems: 1) K^+, $Na^+//Cl^-$, SO_4^{2-}, $Cr_2O_7^{2-}-H_2O$ — solubility isotherm at 100℃; 2) K^+, $Na^+//Cl^-$, SO_4^{2-}, $Cr_2O_7^{2-}-H_2O$ — solubility isotherm at 20℃; 3) $K^+//Cl^-$, SO_4^{2-}, $Cr_2O_7^{2-}-H_2O$ — solubility polythermal (20-100℃); 4) $Na^+//Cl^-$, SO_4^{2-}, $Cr_2O_7^{2-}-H_2O$ — solubility polythermal (20-100℃); 5) K^+, $Na^+//Cr_2O_7^{2-}-H_2O$ — solubility polythermal (20-100℃); 6) K^+, $Na^+//Cl^-$, $SO_4^{2-}-H_2O$ — solubility polythermal (20-100℃).

Let us represent each separately (Figs. 49-54).* A characteristic feature here is that, with the exception of the fifth face, all the faces represent systems with four independent variables and are therefore represented on a plane in the form of optimum projections of triangular prisms. In all cases salt compositions in the range of crystallization of $K_2Cr_2O_7$ are represented.

By arranging the solubility diagrams obtained for these systems on the faces of a cube as shown in Fig. 48, we obtain a model which shows clearly the range of polythermal crystallization of potassium bichromate in the five-component system as a whole, in the temperature range 20-100℃ (Fig. 55).

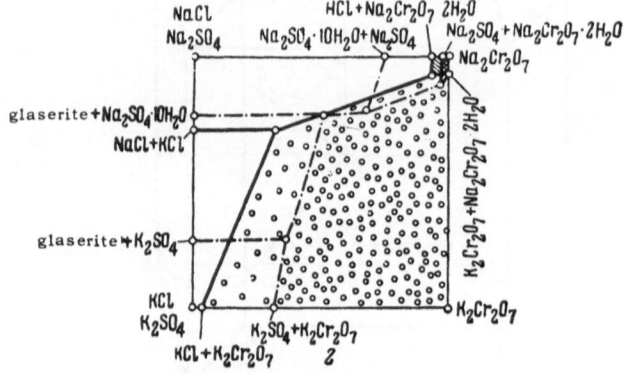

Fig. 50. Solubility isotherm (20℃) for the system Na^+, $K^+//Cl^-$, SO_4^{2-}, $Cr_2O_7^{2-}-H_2O$; salt composition in the field of crystallization of $K_2Cr_2O_7$.

*Constructed using published data [58, 59].

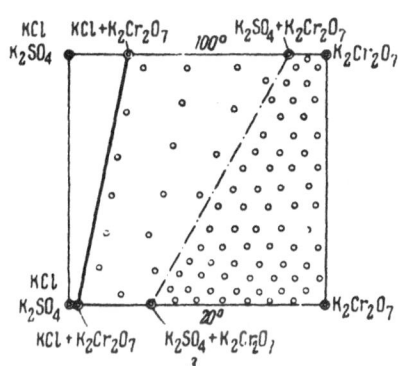

Fig. 51. Solubility polythermal (20-100℃) for the system $K^+//Cl^-$, SO_4^{2-}, $Cr_2O_7^{2-}$—H_2O; salt composition in the field of crystallization of $K_2Cr_2O_7$.

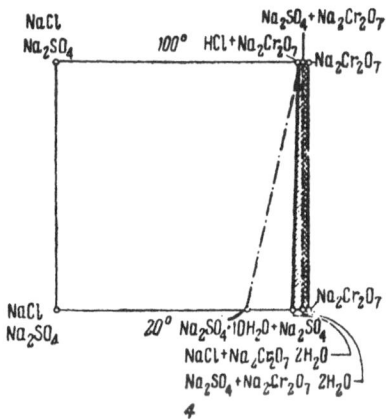

Fig. 52. Solubility polythermal (20-100℃) for the system $Na^+//Cl^-$, SO_4^{2-}, $Cr_2O_7^{2-}$—H_2O; salt composition in the field of crystallization of $K_2Cr_2O_7$.

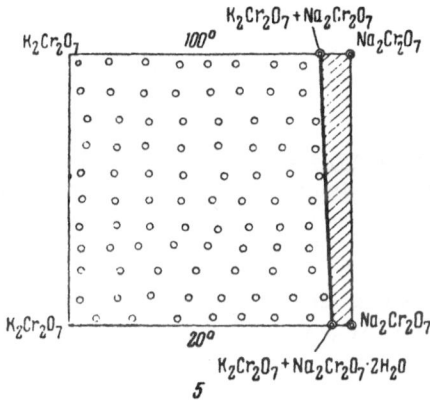

Fig. 53. Solubility polythermal (20-100℃) for the system Na^+, $K^+//Cr_2O_7^{2-}$—H_2O.

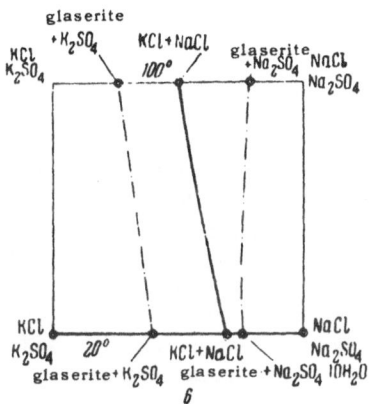

Fig. 54. Solubility polythermal (20-100℃) for the system K^+, $Na^+//Cl^-$, SO_4^{2-}—H_2O; salt composition in the field of crystallization of $K_2Cr_2O_7$.

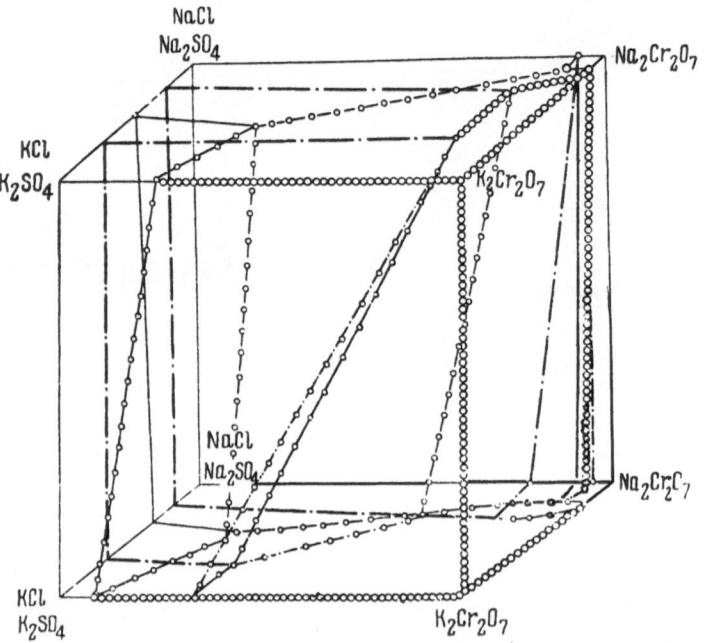

Fig. 55. Model representing the solubility polythermal for the
quinary reciprocal system Na^+, $K^+//Cl^-$, SO_4^{2-}, $Cr_2O_7^{2-}-H_2O$
in the range of crystallization of $K_2Cr_2O_7$.

A characteristic feature of this model and a difference from the usual models for ternary systems lies in
the following. The volume of crystallization of potassium bichromate is here represented by its minimum and
maximum limits: the minimum boundaries of extension of this salt correspond to compositions of the system
with the condition that chlorides are absent; the maximum limits to compositions of the system with the condition that sulfates are absent. From a knowledge of the concentration of chlorides and sulfates for each specific case, we determine the range of extension of the $K_2Cr_2O_7$ fields for both extreme temperatures from
the lever rule.

The boundaries established in this way are of course only tentative, since their determination is based
on the assumption that the temperature dependence of the solubility is linear. Even so, such tentative diagrams, obtained without additional experimental study of isotherms of the system for any intermediate temperatures (between 20 and 100°C) are undoubtedly useful.

To determine the quantity of water corresponding to saturated solutions at both extreme temperatures,
it is necessary to plot on the solubility isotherms (20-100°C) for the five-component system suitable lines corresponding to constant water contents.

LITERATURE CITED

1. A. G. Bergman, Usp. Khim. 5(7-8):1059 (1936).
2. A. G. Bergman and N. S. Dombrovskaya, Izv. Akad. Nauk SSSR, Ser. Khim. No. 1:133 (1936).
3. V. P. Radishchev, Dokl. Akad. Nauk SSSR 21(8):384 (1938).
4. V. P. Radishchev, Dokl. Akad. Nauk SSSR 21(8):393 (1938).
5. V. P. Radishchev, Izv. Sektora Fiz.-Khim. Analiza 22:33 (1953).
6. F. M. Perel'man, Methods for the Representation of Multicomponent Systems. Five-Component Systems, Moscow, Izd. Akad. Nauk SSSR (1959).
7. W. Eitel, Z. Anorg. Chem. 100:95 (1917).
8. V. P. Radishchev, Izv. Sektora Fiz.-Khim. Analiza 14:153 (1941).
9. F. H. C Kelly, J. Chem. Educ. 31(12):637 (1954).
10. F. Flatt, Helv. Chim. Acta 37(1):299 (1954).
11. F. M. Perel'man, Izv. Akad. Nauk SSSR, Otd. Mat. i Estestv. Nauk, No. 2:379 (1936).
12. E. F. Osborn, R. C. de Vries, K. H. Gee, and H. M. Kaner, J. Metals 6(1):33 (1954).
13. H. Klemm, Archiv Metall. 21(7):247 (1949).
14. I. I. Kornilov, Dokl. Akad. Nauk SSSR 81(2):191 (1951).
15. B. N. Delone, Izv. Akad. Nauk SSSR, Otd. Mat. i Estestv. Nauk, No. 1:79 (1929).
16. N. S. Kurnakov, Introduction to Physicochemical Analysis, Moscow-Leningrad, Izd. Akad. Nauk SSSR (1940).
17. V. P. Radishchev, Izv. Sektora Fiz.-Khim. Analiza 23:46 (1953).
18. F. M. Perel'man and A. Ya. Zvorykin, Izv. Sektora Fiz.-Khim. Analiza 26:30 (1955).
19. V. P. Radishchev, Izv. Sektora Fiz.-Khim. Analiza 9:203 (1936).
20. F. M. Perel'man, Zh. Neorgan. Khim. 1:2416 (1956).
21. V. P. Radishchev, Izv. Akad. Nauk SSSR 1:153 (1936).
22. F. M. Perel'man, Zh. Neorgan. Khim. 2:1538 (1957).
23. F. M. Perel'man, Zh. Neorgan. Khim. 3:1611 (1958).
24. F. M. Perel'man, Zh. Neorgan. Khim. 7:900 (1962).
25. F. M. Perel'man, Dokl. Akad. Nauk SSSR 131(3):578 (1960).
26. F. M. Perel'man, Zh. Neorgan. Khim. 5:2007 (1960).
27. F. M. Perel'man, Zh. Neorgan. Khim. 3:630 (1958).
28. F. M. Perel'man, Zh. Neorgan. Khim. 7:844 (1962).
29. I. I. Kornilov and V. S. Vlasov, Zh. Neorgan. Khim. 2:2762 (1957).
30. "Titanium," in collection: Rare Metals [Russian translation], Moscow, IL, 1953-1954.
31. L. Northcott, Metallurgy of the Rarer Metals, Molybdenum, No. 5, London (1956).
32. E. Jänecke, Kurzgefasstes Handbuch aller Legierungen, Heidelberg (1949).
33. P. Pascal, Traité de Chimie Minérale, Vol. 12, Paris (1934).
34. I. I. Kornilov and P. B. Budberg, Zh. Neorgan. Khim. 2:860 (1957).
35. V. S. Mikheev and D. M. Pevtsov, Zh. Neorgan. Khim. 3:861 (1958).
36. V. P. Elyutin and V. P. Funke, Izv. Akad. Nauk SSSR, Otd. Tekhn. Nauk, No. 3, 68 (1956).
37. A. P. Smiryagin, A. Ya. Potemkin, and R. P. Martynov, Zh. Neorgan. Khim. 3:853 (1958).
38. I. I. Kornilov and R. S. Polyakova, Zh. Neorgan. Khim. 3:879 (1958).
39. N. V. Grum-Grzhimailo and D. I. Prokof'ev, Zh. Neorgan. Khim. 3:889 (1958).
40. Yu. A. Bagaryatskii, G. I. Nosova, and T. V. Tagunova, Zh. Neorgan. Khim. 3:777 (1958).
41. Seidell-Linke, Solubilities of Inorganic and Metalorganic Compounds, Vol. 1, 4th ed., New York (1958).
42. F. M. Perel'man, Zh. Neorgan. Khim. 7:896 (1962).

LITERATURE CITED

43. A. Benrath and K. Lechener, Z. Anorg. Chem. 244:359 (1940).
44. V. P. Blidin, Izv. Akad. Nauk SSSR, Otd. Khim. Nauk (1954), p. 409.
45. A. V. Novoselova, R. Danilevich, and A. Tikhonova, Zh. Obshch. Khim. 16:439 (1946).
46. F. A. H. Schreinemakers and W. C. Baat, Z. Phys. Chem. 65:586 (1908-1909).
47. C. Mazetti, Gazz. Chim. Ital. 56:601 (1926).
48. F. A. H. Schreinemakers and W. C. Baat, Chem. Weekblad. 5:465 (1908).
49. V. P. Blidin, Izv. Akad. Nauk SSSR, Otd. Khim. Nauk (1953), p. 814.
50. S. A. Durov, Synthesis in Chemical Hydrology, Rostov-on-Don (1961).
51. F. M. Perel'man, in collection: Science Reviews, Chemical Sciences, Vol. 4, Physicochemical Analysis, Moscow, Izd. Akad. Nauk SSSR (1959), p. 283.
52. N. S. Kurnakov and S. F. Zhemchuzhnyi, Zh. Russ. Fiz.-Khim. Obshchestva 51:1 (1919).
53. E. B. Shternia, Izv. Sektora Fiz.-Khim. Analiza 17:351 (1949).
54. O. K. Yanat'eva, Izv. Sektora Fiz.-Khim. Analiza 20:252 (1954).
55. O. K. Yanat'eva, Dokl. Akad. Nauk SSSR 67:479 (1949).
56. F. K. Kamoron and A. Seidell, J. Phys. Chem. 5:643 (1901).
57. N. K. Voskresenskaya (ed.), Handbook on Fusion in Systems of Anhydrous Inorganic Salts, Vols. 1-2, Moscow-Leningrad, Izd. Akad. Nauk SSSR (1961).
58. A. V. Rakovskii and A. V. Babaeva, Tr. Inst. Chistykh Khim. Reaktivov, No. 11:15 (1931).
59. A. V. Rakovskii and E. A. Nikitina, Tr. Inst. Chistykh Khim. Reaktivov, No. 11:5 (1931).